W9-BGC-972

# *The Air Up There*

# The Air Up There

## More Great Quotations On Flight

Dave English

McGraw-Hill
New York Chicago San Francisco Lisbon London Madrid
Mexico City Milan New Delhi San Juan Seoul
Singapore Sydney Toronto

The McGraw-Hill Companies

Cataloging-in-Publication Data is on file with the Library of Congress.



1 2 3 4 5 6 7 8 9 0 DOC/DOC 0 9 8 7 6 5 4 3

ISBN 0-07-141036-8

*The sponsoring editor for this book was Shelley Carr, the editing supervisor was Steven Melvin, and the production supervisor was Pamela Pelton. It was set in Berkeley by Joanne Morbit of McGraw-Hill Professional's Hightstown, N.J., composition unit.*

*RR Donnelley was printer and binder.*

McGraw-Hill books are available at special quantity discounts to use as premiums and sales promotions, or for use in corporate training programs. For more information, please write to the Director of Special Sales, McGraw-Hill Professional's, Two Penn Plaza, New York, NY 10121-2298. Or contact your local bookstore.

This book is printed on recycled, acid-free paper containing a minimum of 50% recycled, de-inked fiber.

For my Mum, who taught me to read.
And my Dad, who showed me aeroplanes.

# *Contents*

*Introduction* ix
*Magic and Wonder of Flight* 1
*Predictions* 27
*Women Fly* 45
*Combat* 53
*Air Power* 65
*Bums on Seats* 73
*Humor* 81
*Clichés* 91
*Miscellaneous* 97
*Space Flight* 115
*Unidentified Flying Objects* 131
*Skydiving* 137
*Birds* 141
*Ballooning* 147
*Piloting* 153
*Safety* 165
*Last Words* 173
*Index* 177

# Introduction

In late July every year, the center of the flying universe is Oshkosh, Wisconsin. For one magical week the busiest airport in the world is Wittman Regional Airport, home of the Experimental Aircraft Association's AirVenture. There are 12,000 airplanes, air shows every afternoon, acres of exhibitors, forums, speakers, and in the corner of the official gift store something appropriately called the "authors corner." Here aviation writers—both well-known and specialized interest—are available to talk about their books and sign copies. I have been fortunate enough to be a small part of this meetingplace thanks to the success of a book I compiled and edited called *Slipping the Surly Bonds: Great Quotations on Flight*.

One of the surprising questions I was most often asked was, "Will there be a second volume?" After several more years of collecting quotes, this book is the answer to that request. I think it is a good answer. This book contains first-hand lines from the many aviation events—good and bad—that have happened since the first book was published, loads of quotations that have been suggested by readers around the world, and some material that was withheld from the first volume due to space considerations. No one was guessing in 1997 that a book of aviation quotations would sell 30,000 copies or that those 200 pages would not be nearly enough to cover all the material.

If the tidbits here induce an appetite for reading about flying, you cannot go wrong by making a meal of books by Richard Bach, Ernest K. Gann, or Antoine de Saint-Exupéry. Other books that provide a longer sampling of aviation writing from several authors I can happily recommend include *The Greatest Flying Stories Ever Told: Nineteen Amazing Tales from the Sky* (Lamar Underwood, Lyons Press, 2002) and *Wild Blue: Stories of*

*Survival from Air and Space* (David Fisher and William Garvey, Thunder's Mouth Press, 1999). There is also now a definitive book of flying poems, the wonderful *Because I Fly: A Collection of Aviation Poetry* (Helmut H. Reda, McGraw-Hill, 2001). For space enthusiasts the splendid book *Space Shuttle: The First 20 Years—The Astronauts' Experiences in Their Own Words* (Tony Reichhardt, DK Publishing, 2002) contains longer quotations from many astronauts.

This book would not be possible if it were not for the hundreds of e-mails I've received from flying fans around the world. Thank you all for the suggestions and additions to the collection. I would like especially to thank those people who have given me formal permission or personal blessing to reprint their original thoughts. The contact with so many aviation friends has been one of the super experiences I've had since posting a list of first-hand accounts of flight on the Internet many years ago. Excerpts from *Wind, Sand and Stars*, copyright 1939 by Antoine de Saint-Exupéry and renewed 1967 by Lewis Galantiere, are reprinted by kind permission of Harcourt Brace & Company.

If you would like to contact me to suggest more quotations or for any other reason, please visit my website at *www.skygod.com*. Or maybe we will meet one July afternoon, watching airplanes at a small Wisconsin airport . . .

# Magic and Wonder of Flight

The air up there in the clouds is very pure and fine, bracing and delicious. And why shouldn't it be? — it is the same the angels breathe.

—MARK TWAIN
*Roughing It*, Chapter XXII
1886

Man must rise above the Earth — to the top of the atmosphere and beyond — for only thus will he fully understand the world in which he lives.

—SOCRATES

The desire to reach for the sky runs deep in our human psyche.

—CESAR PELLI
ARCHITECT OF THE TALLEST BUILDING IN THE WORLD,
THE TWIN PETRONAS TOWERS IN KUALA LUMPUR.
Quoted in *The New York Times* after the terrorist attack on the World Trade Center
20 September 2001

My soul is in the sky.

—WILLIAM SHAKESPEARE
*A Midsummer Night's Dream*, Act, Scene I

Those who are able to walk on stilts can roam the earth unstopped by mountains or rivers. They are able to imagine flying and therefore reach the isles of the immortals.

—P'AO-POU TSEU

You haven't seen a tree until you've seen its shadow from the sky.

—AMELIA EARHART

Don't let the fear of falling keep you from knowing the joy of flight.

—LANE WALLACE
*Flying* magazine
January 2001

I used to have dreams when I was a kid that I'd go running down the street and jump up in the air and go flying and just fly through the air all by myself. That's what weightlessness is like.

—Robert "Hoot" Gibson
Astronaut

I cannot describe the delight, the wonder and intoxication, of this free diagonal movement onward and upward, or onward and downward. . . . The birds have this sensation when they spread their wings and go tobogganing in curves and spirals through the sky.

—Alberto Santos-Dumont
First dirigible flight

More than anything else the sensation is one of perfect peace mingled with an excitement that strains every nerve to the utmost, if you can conceive of such a combination.

—Wilbur Wright

So let us raise a cheer . . . for the insatiable spirit of Man eager for all new things! What a tale could have been written by that far off man who first saw a tree trunk roll and made a wheel and cart and harnessed in his mare and cracked his whip and drove away to disappear beyond the hill! Or that first man who made a boat and raised a sail and disappeared hull down to unknown shores!

All this is misty in a distant past. The land and sea are long since named and mapped and parcelled out. Only the air and all beyond, the greatest mystery of all, was still unmastered and unknown when I was young. Now we have learned to shuffle about the house and even plan to visit the neighbours. A million starry mansions wink at us as if they knew our hopes and beckon us abroud. All that I shall not see. But at the start, the little lost beginning, I can say of one small part of it: "Here is a witness from my heart and hand and eye of how it was!

—Cecil Lewis
new preface for *Sagittarius Rising* (1936)
1965

By day, or on a cloudless night, a pilot may drink the wine of the gods, but it has an earthly taste; he's a god of the earth, like one of the Grecian deities who lives on worldly mountains and descended for intercourse with men. But at night, over a stratus layer, all sense of the planet may disappear. You know that down below, beneath that heavenly blanket is the earth, factual and hard. But it's an intellectual knowledge; it's a knowledge tucked away in the mind; not a feeling that penetrates the body. And if at times you renounce experience and mind's heavy logic, it seems that the world has rushed along on its orbit, leaving you alone flying above a forgotten cloud bank, somewhere in the solitude of interstellar space.

—CHARLES A. LINDBERGH
*The Spirit of St. Louis*
1961

These bright roofs, these steep towers, these jewel-lakes, these skeins of railroad line—all spoke to her and she answered. She was glad they were there. She belonged to them and they to her. . . . She had not lost it. She was touching it with her fingertips. This was flying: to go swiftly over the earth you loved, touching it lightly with your fingertips, holding the railroads lines in your hand to guide you, like a skein of wool in a spider-web game—like following Ariadne's thread through the Minotaur's maze, Where would it lead, where?

—ANNE MORROW LINDBERGH
*The Steep Ascent*
1944

My father had been opposed to my flying from the first and had never flown himself. However, he had agreed to go up with me at the first opportunity, and one afternoon he climbed into the cockpit and we flew over the Redwood Falls together. From that day on I never heard a word against my flying and he never missed a chance to ride in the plane.

—CHARLES A. LINDBERGH
*We*
1928

Sometimes I watch myself fly. For in the history of human flight it is not yet so very late; and a man may still wonder once in a while and ask: how is it that I, poor earth-habituated animal, can fly?

Any young boy can nowadays explain human flight—mechanistically: " . . . and to climb you shove the throttle all the way forward and pull back just a little on the stick. . . ." One might as well explain music by saying that the further over to the right you hit the piano the higher it will sound. The makings of a flight are not in the levers, wheels, and pedals but in the nervous system of the pilot: physical sensations, bits of textbook, deep-rooted instincts, burnt-child memories of trouble aloft, hangar talk.

—Wolfgang Langewiesche
*A Flyer's World*
1943

I ask people who don't fly, "How can you not fly when you live in a time in history when you can fly?"

—William Langewiesche (Wolfgang's son)
2001

"Just try and remember," I said slowly, "that if God had intended men to fly He'd have given us wings. So all flying is flying in the face of nature. It's unnatural, wicked and stuffed with risks all the time. The secret to flying is learning to minimize the risks."

"Or perhaps—the secret of life is to choose your risks?"

—Gavin Lyall
*Shooting Script*
1966

I think a future flight should include a poet, a priest and a philosopher . . . we might get a much better idea of what we saw.

—Michael Collins
9 November 1969

"You're on your own" was all he said that day long years ago
So long his name and face are lost in memory's afterglow;

Nor do I recollect of pride or joy or doubt or fright
Or other circumstance which marked that time for solo flight
The cryptic words alone endure : he said "you're on your own"
And down through time I've found it so – the test's to walk alone
Not that one choose to draw aside in churlish mein or vein,
From common lot of what life holds of pleasure, toil or pain
But that the call's to rise and cruise alone with dreams unshared
Or plan alone for some far goal, for which none else has cared
Or fight alone for what you hold is worth a warrior's strife

And ask no gain or fame or aught beyond the joy of life.
I owe a quenchless debt to him who bade me seize my fate
And hang it on the faith that I to it was adequate
For when he said "you're on your own" and sauntered on away
I knew that here, in four short words, was youth's first judgement day
Not wit to learn, not test of skill, not pride to satisfy,
But will to walk down life in faith that life is theirs who try.

—Gill Robb Wilson
*Flying* magazine
January 1961

I will fly in the greatness of God as the marsh-hen flies,
In the freedom that fills all the space 'twixt the marsh and the skies.

—Sidney Lanier
American poet
in the poem "The Marshes of Glynn"

We who fly are going to get to know that Great Flying Boss in the sky better and better.

—Colonel Robert L. Scott, Jr., USAAF
*God Is My Copilot*
1943

Feathers shall raise men even as they do birds, toward heaven; that is by letter written with their quills.

—Leonardo da Vinci

As you pass from sunlight into darkness and back again every hour and a half, you become startingly aware how artificial are thousands of boundaries we've created to separate and define. And for the first time in your life you feel in your gut the precious unity of the Earth and all the living things it supports.

—RUSSELL SCHWEICKART
**Returning from Apollo 9**

Nowadays a businessman can go from his office straight to the airport, get into his airplane and fly six hundred or seven hundred miles without taking off his hat. He probably will not even mention this flight, which a bare twenty-five years ago would have meant wearing leather jacket and helmet and goggles and risking his neck every minute of the way.

No, he probably wouldn't mention it - except to another flier. Then they will talk for hours. They will re-create all the things seen and felt in that wonderful world of air: the sense of remoteness from the busy world below, the feeling of intense brotherhood formed with those who man the radio ranges and control towers and weather stations that bring the pilot home, the clouds and the colors, the surge of the wind on their wings.

They will speak of things that are spiritual and beautiful and of things that are practical and utilitarian; they will mix up angels and engines, sunsets and spark plugs, fraternity and frequencies in one all-encompassing comradeship of interests that makes for the best and most lasting kind of friendship any man can have.

—PERCY KNAUTH
***Wind on My Wings***
**1960**

This was the crystalline moment Dan loved so well, the moment of transition between ground and air, when the laws of aerodynamics took over the job of physical support of the jet. He'd become a pilot for this very moment: the feel of mighty engines and the roar of the slipstream, all converging on the reality of sustained flight on an invisible highway of air. Flying was a thrill in even a single-engine airplane, but to levitate a leviathan—a metallic eggshell longer than a football field and heavier than a house—was a magic he could never quite comprehend. Every liftoff was a philosophical wonder that left a broad smile on his face.

—JOHN J. NANCE
*Blackout*
2000

And if flying, like a glass-bottomed bucket, can give you that vision, that seeing eye, which peers down on the still world below the choppy waves—it will always remain magic.

—ANNE MORROW LINDBERGH
*North to the Orient*
1935

More varied than any landscape was the landscape in the sky, with islands of gold and silver, peninsulas of apricot and rose against a background of many shades of turquoise and azure.

—CECIL BEATON
quoted in *Cecil Beaton*
Hugo Vickers
*regards an Egyption sunset*
1985

The air is the most mysterious, the most exciting, the most challenging of all the elements. We leave the planet, we leave the sea, we leave the earth. The air is no longer of this world . . .

—DAVID BEATY

As we got further and further away, it [the Earth] dimished in size. Finally it shrank to the size of a marble, the most beautiful you can imagine. That beautiful, warm, living object looked so fragile, so delicate, that if you touched it with a finger it would crumble and fall apart. Seeing this has to change a man.

—JAMES B. IRWIN
Apollo 15

To see the earth as it truly is, small and blue and beautiful in that eternal silence where it floats, is to see ourselves a riders on the earth together, brothers on that bright loveliness in the eternal cold—brothers who know now they are truly brothers.

—ARCHIBALD MACLEISH

The last of the lonely places is the sky, a trackless void where nothing lives or grows, and above it, space itself. Man may have been destined to walk upon ice or sand, or climb the mountains or take craft upon the sea. But surely he was never meant to fly? But he does, and finding out how to do it was his last great adventure.

—Frederick Forsyth

My senses of space, of distance, and of direction entirely vanished. When I looked for the ground I sometimes looked down, sometimes up, sometimes left, sometimes right. I thought I was very high up when I would suddenly be thown to earth in a near vertical spin. I thought I was very low to the ground and I was pulled up to 3,000 feet in two minutes by the 500-horsepower motor. It danced, it pushed, it tossed. . . . Ah! la la!

—Antoine de Saint-Exupéry
***Lettres à sa Mère***
*letter to his mother regards his first flight in a SPAD-Herbemont.*
*This was one of his first flights, and these are his first words on the experience of flight*
**1921**

Splutter, splutter. Yes—we're off—we're rising. But why start off with an engine like that? But it smooths out now, like a long sigh, like a person breathing easily, freely. Like someone singing ecstatically, climbing, soaring - sustained note of power and joy. We turn from the lights of the city; we pivot on a dark wing; we roar over the earth. The plane seems exultant now, even arrogant. We did it, we did it! We're up, above you. We were dependant on you just now, prisoners fawning on you for favors, for wind and light. But now, we are free. We are up; we are off. We can toss you aside, for we are above it.

—Anne Morrow Lindbergh
***Listen! the Wind***
**1938**

Ours is the commencement of a flying age, and I am happy to have popped into existence at a period so interesting.

—Amelia Earhart
***20 Hrs 40 Mins***
**1928**

Possibly everyone will travel by air in another fifty years. I'm not sure I like the idea of millions of planes flying around overhead. I love the sky's unbroken solitude. I don't like to think of it cluttered up by aircraft, as roads are cluttered up by cars. I feel like the western pioneer when he saw barbed-wire fence lines encroaching on his open plains. The success of his venture brought the end of the life he loved.

—CHARLES A. LINDBERGH
*The Spirit of St. Louis*
1953

All was glorious—a cloudless sky above, a most delicious view around. . . . How great is our good fortune! I care not what may be the condition of the earth; it is the sky that is for me now.

—PROFESSOR JACQUES ALEXANDRE CESARE CHARLES
*first free flight in a manned hydrogen balloon*
December 1, 1783

To confine our attention to terrestrial matters would be to limit the human spirit.

—STEPHEN HAWKING

Flying was a very tangible freedom. In those days, it was beauty, adventure, discovery—the epitome of breaking into new worlds.

—ANNE MORROW LINDBERGH
introduction to *Hour of Gold, Hour of Lead*
1929

It was a magic caused by the collision of modern methods and old ones; modern history and ancient; accessibility and isolation. And it was a magic which could only strike spark about that time. A few years earlier, from the point of view of aircraft alone, it would have been impossible to reach these places; a few later, and there will be no such isolation.

—ANNE MORROW LINDBERGH
the preface to *North to the Orient*
1935

Beyond the Sudd there is the desert, and nothing but the desert for almost three thousand miles, nor are the towns and cities that live in it succesful in gainsaying its emptiness.

To me, desert has the quality of darkness; none of the shapes you see in it are real or permanent. Like night, the desert is boundless, comfortless, and infinite. Like night, it intrigues the mind and leads it to futility. When you have flown halfway across a desert, you experience the desperation of a sleepless man waiting for dawn which only comes when the importance of its coming is lost.

—Beryl Markham
*West with the Night*
1942

Using an artful tool does not make one a dry technician. It seems to me that people that are anxious about our technical advancement, confuse means and ends. Naturally a person that only works for material gain will not harvest something that is worth living for. But the machine is not an end in itself. The airplane is not an end. It is a tool. Just like the plough.When we think that the machine will harm man, then it is perhaps because we are not yet capable of judging the rapid changes it has brought about. We hardly feel at home in this landscape of mines and power stations. We have just moved into this new home that we have not even finished yet. Everything around us has changed so fast - personal relations, working conditions, habits. Even our state of mind is in turmoil.

We are all youthful barbarians, and only our new toys bring us excitement. That has been the sole purpose of our flights. This one flies higher, that one faster. But now we will make ourselves at home. We will forget the machine, the tool. It is no longer complex; it does what it is supposed to do, unnoticed.

And through this tool we will find again the old nature, the nature of the gardener, the navigator, the poet.

—Antoine de Saint-Exupéry
*Wind, Sand, and Stars*
1939

The philosopher is Nature's pilot. And there you have our difference: to be in hell is to drift: to be in heaven is to steer.

—George Bernard Shaw

*BECAUSE I FLY*

Because I fly
I laugh more than other men
I look up and see more than they,
I know how the clouds feel,
What it's like to have the blue in my lap,
to look down on birds,
to feel freedom in a thing called the stick...
who but I can slice between God's billowed legs,
and feel then laugh and crash with His step
Who else has seen the unclimbed peaks?
The rainbow's secret?
The real reason birds sing?
Because I Fly,
I envy no man on earth.

—Anonymous

I can't remember the time when airplanes were not a part of my life and can't remember ever wanting anything so much as to fly one. Once I had started I had to keep flying.

But it was not until I was seventeen that I finally got into an airplane. At that time I felt I had come to the place where I belonged in the world. The air to me was what being on the ground was to other people. When I felt nervous it pulled me together. Things could get too much for me on the ground, they never got that way in the air. Flying came into my mind like fresh air into smoked up lungs and was food in my hungry mouth and strength in my weak arms. I felt that way the first time I got into an airplane. I wasn't nervous when I first soloed. There was excitement in me, but it was the nice kind you get when you're going home after a long, long unhappy time away.

—Major Don S. Gentile
USAAF

I had that morning gone to say my farewells to Broadhurst and to the RAF. I had made a point of going to HQ at Schleswig in my 'Grand Charles'. Coming back I had taken him high up in the cloudless summer sky, for it was only there that I could fittingly take my leave.

Together we climbed for the last time straight towards the sun. We looped once, perhaps twice, we lovingly did a few slow, meticulous rolls, so that I could take away in my finger-tips the vibration of his supple, docile wings.

And in that narrow cockpit I wept, as I shall never weep again, when I felt the concrete brush against his wheels and, with a great sweep of the wrist, dropped him on the ground like a cut flower.

As always, I carefully cleared the engine, turned off all the switches one by one, removed the straps, the wires and the tubes which tied me to him, like a child to his mother. And when my waiting pilots and my mechanics saw my downcast eyes and my shaking shoulders, they understood and returned to the dispersal in silence.

—PIERRE CLOSTERMANN
*The Big Show*
1951

In the case of pilots, it is a little touch of madness that drive us to go beyond all known bounds. Any search into the unknown is an incomparable exploitation of oneself.

—JACQUELINE AURIOL

Where did you get your eyes so blue?
Out of the sky as I came through.

—GEORGE MACDONALD
*At the Back of the North Wind*

Every flyer who ventures across oceans to distant lands is a potential explorer; in his or her breast burns the same fire that urged adventurers of old to set forth in their sailing-ships for foreign lands. . . . Riding through the air on silver wings instead of sailing the seas with white wings, he must steer his own course, for the air is uncharted, and he must therefore explore for himself the strange eddies and currents of the ever-changing sky in its many moods.

—JEAN BATTEN
*Alone in the Sky*
1979

I had always wanted an adventurous life. It took a long time to realize that I was the only one who was going to make an adventurous life happen to me.

—RICHARD BACH
Interview

The exhilaration of flying is too keen, the pleasure too great, for it to be neglected as a sport.

—ORVILLE WRIGHT

Flying is more than a sport and more than a job; flying is pure passion and desire, which fill a lifetime.

—GENERAL ADOLF GALLAND, LUFTWAFFE
*The First and the Last*
1954

I am with the angels and just completely happy.

—BERTRAND PICCARD
*Swiss pilot of Breitling Orbiter 3, first to balloon around the world*
20 March 1999

I am going to have a cup of tea, like any good Englishman.

—BRIAN JONES
*British pilot of Breitling Orbiter 3, first to balloon around the world*
20 March 1999

I live for that exhilarating moment when I'm in an airplane rushing down the runway and pull on the stick and feel lift under its wings. It's a magical feeling to climb toward the heavens, seeing objects and people on the ground grow smaller and more insignificant. You have left that world beneath you. You are inside the sky.

—L. GORDON 'GORDO' COOPER
*Leap of Faith*
2000

I'll run my hand gently over the wing of a small airplane and say to him, "This plane can teach you more things and give you more gifts than I ever could. It won't get you a better job, a faster car, or a bigger house. But if you treat it with respect and keep your eyes open, it may remind you of some things you used to know—that life is in the moment, joy matters more than money, the world is a beautiful place, and that dreams really, truly are possible." And then, because airplanes speak in a language beyond words, I'll take him up in the evening summer sky and let the airplane show him what I mean.

—LANE WALLACE
"Eyes of a Child"
*Flying* magazine
February 2000

Before I went to the Mess I made the excuse I wanted to get something out of my aeroplane, and climbed into the cockpit; I did this, however, to be able to say good-bye to the old dear; and I really felt dreadfully sorry to part with her. I get very attached to aeroplanes, and I am one of those people who think that they aren't so inanimate as we are told they are.

—CHARLES RUMNEY SAMSON
*A Flight from Cairo to Cape Town and Back*
1931

Ah hell. We had more fun in a week than those weenies had in a lifetime.

—PANCHO BARNES
quoted in *The Happy Bottom Riding Club: The Life and Times of Pancho Barnes*
Lauren Kesler
2000

Here above the farms and ranches of the Great Plains aviation lives up to the promise that inspired dreamers through the ages. Here you are truly separate from the earth, at least for a little while, removed from the cares and concerns that occupy you on the ground. This separation from the earth is more than symbolic, more than a physical removal - it has an emotional dimension as tangible as the wood, fabric, and steel that has transported you aloft.

—Stephen Coonts
*The Cannibal Queen*
1988

On the second day after I arrived at Cranwell I was commanded to report to 'the flights'. I had imagined weeks if not months of tedious "bull" and ground instruction before I was even allowed to smell an aircraft.

They had a special smell - burnt castor oil and dope - which will still bring nostalgic sparkles to the eyes of an old pilot.

—A.G. Dudgeon
*The Luck of the Devil*
1985

The engine is the heart of an aeroplane, but the pilot is its soul.

—Sir Walter Alexander Raleigh
(circa 1910, not of course the Sir Walter Raleigh beheaded some 300 years earlier)

Be like the bird in flight . . . pausing a while on boughs too slight, feels them give way beneath her, yet sings knowing yet, that she has wings.

—Victor Hugo

We are all in the gutter but some of us are looking at the stars.

—Oscar Wilde

For my part, I travel not to go anywhere, but to go. I travel for travel's sake. The great affair is to move.

—Robert Louis Stevenson
*Travels with a Donkey*
1879

Are we lost, or are we found at last?

On earth we strive for our various needs, because so goes the fundamental law of man. Aloft, at least for a little while, the needs disappear. Likewise the striving.

In the thoughts of man aloft, food and evil become mixed and sometimes reversed. This is the open door to wisdom.

Aloft, the earth is ancient and man is young, regardless of his numbers, for there, aloft he may reaffirm his suspicions that he may not be so very much. This is the gateway to humility.

And yet, aloft there are moments when man can ask himself, "what am I, this creature so important to me? Who is it rules me from birth to tomb? Am I but a slave destined to crawl for labor to hearth and back again? Am I but one of the living dead, or my own god set free?" This is the invitation to full life. . . .

"Where are we?"

"If you really must know, I'll tell you."

"Never mind. Here aloft, we are not lost, but found."

—Ernest K. Gann
*Ernest K. Gann's Flying Circus*
1974

I might have been born in a hovel, but I determined to travel with the wind and the stars.

—Jackie Cochran

Courage is the price that life extracts for granting peace. The soul that knows it not, knows no release from little things.

The soul that knows it not knows no release from little things.
Knows not the livid loneliness of fear,
Nor mountain heights, where bitter joy can hear
The sound of wings.

—Amelia Earhart

*ONE MORE ROLL*

We toast our hearty comrades who have fallen from the skies, and were gently caught by God's own hand to be with him on High.
To dwell among the soaring clouds they've known so well before. From victory roll to tail chase, at heaven's very door.
As we fly among them there, we're sure to head their plea. To take care my friend, watch your six, and do one more roll for me.

—COMMANDER JERRY COFFEE
*Hanoi*
1968

You are brave. Not brave because you are going to be facing any physical dangers; you are not really going to. I mean brave in another, deeper sense. By being on this flight you have shown that you are willing to explore your own identity to discover what might lie within you. Your human clay has not hardened, and you are also willing to explore your own perceptions of the universe, knowing that you may be forced to set aside many comfortable and cherished assumptions. The idea that you must approach honestly and directly is that flying very dramatically makes the pilot solely responsible for his own life.

—HARRY BAUER
*The Flying Mystique: Exploring Reality and Self in the Sky*
1980

It suddenly struck me that that tiny pea, pretty and blue, was the Earth. I put up my thumb and shut one eye, and my thumb blotted out the planet Earth. I didn't feel like a giant. I felt very, very small.

—NEIL ARMSTRONG
Apollo 11

I may be flying a complicated airplane, rushing through space, but in this cabin I'm surrounded by simplicity and thoughts set free of time. How detached the intimate things around me seem from the great world down below. How strange is this combination of proximity and separation. That ground—seconds away—thousands of miles away. This air, stirring mildly around me. That air, rushing by with the speed of a tornado, an inch beyond. These minute details in my cockpit. The grandeur of the world outside. The nearness of death. The longness of life.

—CHARLES A. LINDBERGH
*The Spirit of St. Louis*
1953

Looking at the stars always makes me dream, as simply as I dream over the black dots representing towns and villages on a map. Why, I ask myself, shouldn't the shining dots of the sky be as accessible as the black dots on the map of France?

—VINCENT VAN GOGH

It may be that the invention of the aeroplane flying-machine will be deemed to have been of less material value to the world than the discovery of Bessemer and open-hearth steel, or the perfection of the telegraph, or the introduction of new and more scientific methods in the management of our great industrial works. To us, however, the conquest of the air, to use a hackneyed phrase, is a technical triumph so dramatic and so amazing that it overshadows in importance every feat that the inventor has accomplished. If we are apt to lose our sense of proportion, it is not only because it was but yesterday that we learned the secret of the bird, but also because we have dreamed of flying long before we succeeded in ploughing the water in a dug-out canoe. From Icarus to the Wright Brothers is a far cry.

—WALDEMAR KAEMPFFERT
*The New Art of Flying*
1910

It looked like a fiery sword going into the sky. There came this enormous roar and the whole sky seemed to vibrate; this kind of unearthly roaring was something human ears had never heard. It is very hard to describe what you feel when you stand on the threshold of a whole new era; of a whole new age. . . . It's like those people must have felt—Columbus or Magellan—that for the first time saw entire new worlds and knew the world would never be the same after this. . . . We know the space age had begun.

—Dr. Walter Robert Dornberger
*regards the first successful flight of the A-4 rocket, which went to the edge of space*
3 October 1942

Frequently on the lunar surface I said to myself, 'This is the Moon, that is the Earth. I'm really here, I'm really here!'

—Alan Bean
Apollo 12

We were flying over America and suddenly I saw snow, the first snow we ever saw from orbit. I have never visited America, but I imagined that the arrival of autumn and winter is the same there as in other places, and the process of getting ready for them is the same. And then it struck me that we are all children of our Earth.

—Aleksandr Aleksandrov

The human space program has existed in the collective unconscious of humanity since the dawn of awareness.

—Frank White
*The Overview Effect: Space Exploration and Human Evolution*
1987

Through you, we feel as giants, once again.

—President Ronald Reagan
*to the crew of Columbia after their completion of the first shuttle mission*
14 April 1981

I owned the world that hour as I rode over it. . . . free of the earth, free of the mountains, free of the clouds, but how inseparably I was bound to them.

—Charles A. Lindbergh
quoted in *Lindbergh*
Leonard Mosley
*on flying above the Rocky Mountains*
1978

And should I not, had I but known, have flung the machine this way and that, once more to feel it live under my hand, have sported in the sky and laughed and sung, knowing that never after should I feel so free, so sure in hazard, so secure, riding the daylight in the pride of youth? No more horizons wider than Hope! No more the franchise of the sky, the freedom of the blue! No more! Farewell to wings! Down to the little earth!

That distant day had a significance I could not give it then. So we wheeled and came back south towards the city. The Temple of Heaven slipped by underneath, that perfect pattern in its ample park. Then the wide plain ruled to the far horizon. Soon the aerodrome.

Now shut the engines off. Come down and flatten out, feel the long float, and at the given moment pull the stick right home. She's down. Now taxi in. Switch off. It's over - but not quite, for the port engine, just as if it knew, as if reluctant at the last to let me go, kicked, kicked, and kicked again, as overheated engines will, then backfired with an angry snorting: Fool! The best is over . . .But I did not hear.

—Cecil Lewis
*Sagittarius Rising*
*regards flying for the last time a Vickers Vimy over Peking, 1921*
1936

The sea is dangerous and its storms terrible, but these obstacles have never been sufficient reason to remain ashore. . . . Unlike the mediocre, intrepid spirits seek victory over those things that seem impossible. . . . It is with an iron will that they embark on the most daring of all endeavors. . . . to meet the shadowy future without fear and conquer the unknown.

—Ferdinand Magellan
circa 1520

One cannot look at the sea without wishing for the wings of a swallow.

—Sir Richard Burton

The magic of the craft has opened for me a world in which I shall confront, within two hours, the black dragons and the crowned crests of a coma of blue lightnings, and when night has fallen I, delivered, shall read my course in the starts.

—ANTOINE DE SAINT-EXUPÉRY
*Wind, Sand, and Stars*
1939

So the crew fly on with no thought that they are in motion. Like night over the sea, they are very far from the earth, from towns, from trees. The clock ticks on. The dials, the radio lamps, the various hands and needles go though their invisible alchemy. . . . and when the hour is at hand the pilot may glue his forehead to the window with perfect assurance. Out of oblivion the gold has been smelted: there it gleams in the lights of the airport.

—ANTOINE DE SAINT-EXUPÉRY
*Wind, Sand, and Stars*
1939

When I was twenty, most of my friends were dead. We had sweated out the troopship journey together, shared the excitements of new countries, endured and enjoyed the efforts of learning to fly. At last we had completed our training, and had stood in the hot Rhodesian sun together while our wings were pinned on our chests. We were then more than friends; we were fellow pilots, which to a boy of nineteen was inexpressibly wonderful. . .

—CAPTAIN LINCOLN LEE
first words of *Three-Dimensioned Darkness: The World of the Airline Pilot*
1962

Father, we thank you, especially for letting me fly this flight … for the privilege of being able to be in this position, to be in this wondrous place, seeing all these many startling, wonderful things that you have created.

—L. GORDON "GORDO" COOPER
quoted in the *New York Times*
*prayer while orbiting the earth*
22 May 1963

The facts are that flying satisfies deeply rooted desires. For as long as time these desires have hungered vainly for fulfillment. The horse, and later the motorcar, have merely teased them. The upward sweep of the airplane signifies release.

—Bruce Gould
*Sky Larking*
1929

Birds in flight, claims the architect Vincenzo Volentieri, are not between places—they carry their places with them. We never wonder where they live: they are at home in the sky, in flight. Flight is their way of being in the world.

—Geoff Dyer

The job has its grandeurs, yes. There is the exultation of arriving safely after a storm, the joy of gliding down out of the darkness of night or tempest toward a sun-drenched Alicante or Santiago; there is the swelling sense of returning to repossess one's place in life, in the miraculous garden of earth, where are trees and women and, down by the harbor, friendly little bars. When he has throttled his engine and is banking into the airport, leaving the somber cloud masses behind, what pilot does not break into song?

—Antoine de Saint-Exupéry
*Night Flight*
1933

Any pilot can describe the mechanics of flying. What it can do for the spirit of man is beyond description.

—U.S. Senator Barry M. Goldwater

And there is a Catskill eagle in some souls that can alike dive down into the blackest gorges, and soar out of them again and become invisible in the sunny spaces. And even if he for ever flies within the gorge, that gorge is in the mountains; so that even in his lowest swoop the mountain eagle is still higher than other birds upon the plain, even though they soar.

—Herman Melville
*Moby-Dick, or, the Whale*, Chapter 96
1851

Pilots track their lives by the number of hours in the air, as if any other kind of time isn't worth noting.

—MICHAEL PARFIT
**"The Corn Was Two Feet Below the Wheels"**
***Smithsonian* magazine, May 2000**

Aviators live by hours, not by days.

—T. H. WHITE
***England Have My Bones***
**1936**

I don't understand them anymore, these people that travel the commuter-trains to their dormitory towns. These people call themselves human, but, by a pressure they do not feel, are forced to do their work like ants. With what do they fill their time when they are free of work on their silly little Sundays?

I am very fortunate in my profession. I feel like a farmer, with the airstrips as my fields. Those that have once tasted this kind of fare will not forget it, ever. Not so, my friends? It is not a question of living dangerously. That formula is too arrogant, too presumptuous. I don't care much for bull-fighters. It's not the danger I love. I know what I love. It is life itself.

—ANTOINE DE SAINT-EXUPÉRY
***Wind, Sand, and Stars***
**1939**

I think it is a pity to lose the romantic side of flying and simply to accept it as a common means of transport, although that end is what we have all ostensibly been striving to attain.

—AMY JOHNSON
***Sky Roads of the World***
**1939**

I would recommend a solo flight to all prospective suicides. It tends to make clear the issue of whether one enjoys being alive or not.

—T. H. WHITE
***England Have My Bones***
**1936**

Within all of us is a varying amount of space lint and star dust, the residue from our creation. Most are too busy to notice it, and it is stronger in some than others. It is strongest in those of us who fly and is responsible for an unconscious, subtle desire to slip into some wings and try for the elusive boundaries of our origin.

—K. O. Eckland
*Footprints on Clouds*
1976

Soon there will be no one who remembers when spaceflight was still a dream, the reverie of reclusive boys and the vision of a handful of men.

—Wyn Wachhorst
1995

Never stop being a kid. Never stop feeling and seeing and being excited with great things like air and engines and sounds of sunlight within you. Wear your little mask if you must to protect you from the world but if you let that kid disappear you are grown up and you are dead.

—Richard Bach
*Nothing by Chance*
1969

His talent was as natural as the pattern that was made by the dust on a butterfly's wings. At one time he understood it no more than the butterfly did and he did not know when it was brushed or marred. Later he became conscious of his damaged wings and of their construction and he learned to think and could not fly any more because the love of flight was gone and he could only remember when it had been effortless.

—Ernest Hemingway
*A Moveable Feast*
1964

The highest art form of all is a human being in control of himself and his airplane in flight, urging the spirit of a machine to match his own.

—Richard Bach
*A Gift of Wings*
1974

# Predictions

Engines of war have long since reached their limits, and I see no further hope of any improvement in the art.

—FRONTINUS, A.D. 90

Who would have believed that a huge ocean could be crossed more peacefully and safely than the narrow expanse of the Adriatic, the Baltic Sea or the English Channel? Provide ship or sails adapted to the heavenly breezes, and there will be some who will not fear even that void [of space]. . . . So, for those who will come shortly to attempt this journey, let us establish the astronomy: Galileo, you of Jupiter, I of the moon.

—JOHANNES KEPLER
letter to Galileo
*Conversation with the Messenger from the Stars*
19 April 1610

Yet I do seriously and on good grounds affirm it possible to make a flying chariot in which a man may sit and give such a motion unto it as shall convey him through the air. And this perhaps might be made large enough to carry divers men at the same time, together with food for their viaticum and commodities for traffic. It is not the bigness of anything in this kind that can hinder its motion, if the motive faculty be answerable thereunto. We see a great ship swims as well as a small cork, and an eagle flies in the air as well as a little gnat. . . . 'Tis likely enough that there may be means invented of journeying to the moon; and how happy they shall be that are first successful in this attempt.

—JOHN WILKINS
*A Discourse Concerning a New World and Another Planet*, Book 1
1640

A time will come when men will stretch out their eyes. They should see planets like our Earth.

—CHRISTOPHER WREN
1657

I believe I have found a way to make a machine lighter than air itself. . . . We live submerged at the bottom of an ocean of the element air, which by unquestioned experiments is known to have weight.

—Francesco de Lana de Terzi of Brescia
*the first to propose a flying machine based on sound scientific principles*
1670

The time will come, when thou shalt lift thine eyes
To watch a long-drawn battle in the skies.
While aged peasants, too amazed for words,
Stare at the flying fleets of wondrous birds.
England, so long mistress of the sea,
Where winds and waves confess her sovereignty,
Her ancient triumphs yet on high shall bear
And reign the sovereign of the conquered air.

—Thomas Gray
1737

I am well convinced that "Aerial Navigation" will form a most prominent feature in the progress of civilisation.

—Sir George Cayley
1804

You would make a ship sail against the winds and currents by lighting a bonfire under her deck . . . I have no time for such nonsense.

—Napoleon Bonaparte
*regards Fulton's Steamship*

Railroad carriages are pulled at the enormous speed of fifteen miles per hour by engines which, in addition to endangering life and limb of passengers, roar and snort their way through the countryside, setting fire to the crops, scaring the livestock, and frightening women and children. The Almighty certainly never intended that people should travel at such break-neck speed.

—President Martin van Buren
1829

Of all inventions of which it is possible to conceive in the future, there is none which so captivates the imagination as that of a flying machine. The power of rising up into the air and rushing in any direction desired at the rate of a mile or more in a minute is a power for which mankind would be willing to pay very liberally. What a luxurious mode of locomotion! To sweep along smoothly, gracefully, and swiftly over the treetops, changing course at pleasure, and alighting at will. How perfectly it would eclipse all other means of travel by land and sea! This magnificent problem, so alluring to the imagination and of the highest practical convenience and value, has been left heretofore to the dreams of a few visionaries and the feeble efforts of a few clumsy inventors. We, ourselves, have thought that, in the present state of human knowledge, it contained no promise of success. But, considering the greatness of the prize and the trifling character of the endeavors which have been put forth to obtain it, would it not indeed be well, as our correspondents suggest, to make a new and combined effort to realize it, under all the light and power of modern science and mechanism? . . .

The simplest, however, of all conceivable flying machines would be a cylinder blowing out gas in the rear and driving itself along on the principle of the rocket. . . .

We might add several other hints to inventors who desire to enter on this enticing field, but we will conclude with only one more. The newly discovered metal aluminum, from its extraordinary combination of lightness and strength, is the proper material for flying machines.

—*Scientific American*
**8 September 1860**

In spite of the opinions of certain narrow-minded people who would shut up the human race upon this globe, we shall one day travel to the moon, the planets, and the stars with the same facility, rapidity and certainty as we now make the ocean voyage from Liverpool to New York.

—Jules Verne
1865

I believe, sir, in all the progress. Air navigation is the result of the oceanic navigation: from water the human has to pass in the air. Everywhere where creation will be breathable to him, the human will penetrate into the creation. Our only limit is life. There where ends the air column which prevents our machine to burst, the human has to stop. But he can, owes, and wants to go to there, and he will go. You can do it. I take the biggest interest in your useful and brave perpendicular journeys. Your ingenious and fearless companion, Mr W. de Fontevielle, has as Mr. Victor Meunier the superior instinct of the true science. I would have the magnificent taste of the scientific adventure. Adventure in the fact, the hypothesis in the idea, here is the two big processes of discovery. Certainly, the future is for air navigation and the duty of the present is to work for the future. You are just now endorsing this duty. I, solitary person, but attentive, I am your eyes and I say to you: Courage!

—VICTOR HUGO
**letter sent to Gaston Tissandier**
**9 March 1869**

To set foot on the soil of the asteroids, to lift by hand a rock from the Moon, to observe Mars from a distance of several tens of kilometers, to land on its satellite or even on its surface, what can be more fantastic? From the moment of using rocket devices a new great era will begin in astronomy: the epoch of the more intensive study of the firmament.

—KONSTANTIN E. TSIOLKOVSKY

Consider a cask filled with a highly compressed gas. If we open one of its taps the gas will escape through it in a continuous flow, the elasticity of the gas pushing its particles into space will continuously push the cask itself. The result will be a continuous change in the motion of the cask. Given a sufficient number of taps (say, six), we would be able to regulate the outflow of the gas as we liked and the cask (or sphere) would describe any curved line in accordance with any law of velocities.

—KONSTANTIN E. TSIOKOVSKY
*regards how a rocket works in space*
*Free Space*
1883

One: There is a low limit of weight [of about] 50 pounds beyond which it is impossible for an animal to fly. Two: The animal machine is far more effective than any we can hope to make. Three: The weight of any machine constructed for flying, including fuel and engineer, cannot be less than three or four hundred pounds. Is it not demonstrated that a true flying machine, self-raising, self-sustaining, self-propelling, is physically impossible?

—PROFESSOR JOSEPH LE CONTE
**University of California**
**1888**

It is apparent to me that the possibilities of the aeroplane, which two or three years ago were thought to hold the solution to the [flying machine] problem, have been exhausted, and that we must turn elsewhere.

—THOMAS EDISON
**1895**

I have not the smallest molecule of faith in aerial navigation other than ballooning, or of the expectation of good results from any of the trials we heard of. So you will understand why I would not care to be a member of your society.

—LORD KELVIN
**replying to an invitation to join the Royal Aeronautical Society**
**1896**

The energy necessary to propel a ship would be many times greater than that required to drive a train of cars at the same speed; hence as a means of rapid transit, flying could not begin to compete with the railroad.

—*POPULAR SCIENCE* MAGAZINE
**1897**

Aerial flight is one of that class of problems with which man will never be able to cope.

—SIMON NEWCOMB
**circa 1900**

All attempts at artificial aviation are not only dangerous to life but doomed to failure from an engineering standpoint.

—THE EDITOR OF *THE TIMES* OF LONDON
1905

It is complete nonsense to believe flying machines will ever work.

—SIR STANLEY MOSLEY
1905

Their Lordships are of the opinion that they would not be of any practical use to the Naval Service.

—BRITISH ADMIRALTY
in reply to the Wright's offer of patents for their airplane
1907

It is a bare possibility that a one-man machine without a float and favored by a wind of, say, 15 miles an hour, might succeed in getting across the Atlantic. But such an attempt would be the height of folly. When one comes to increase the size of the craft, the possibility rapidly fades away. This is because of the difficulties of carrying sufficient fuel. It will readily be seen, therefore, why the Atlantic flight is out of the question.

—ORVILLE WRIGHT
circa 1908

It will take much longer [than the automobile] to make them [airplanes] familiar to everyone, yet nobody should lose sight of the fact that the Age of Flight is really here, that the man-bird is fledged at last, and already on the wing.

—EDITORIAL IN *OUTING*
1909

We soon saw that the helicopter had no future, and dropped it. The helicopter does with great labor only what the balloon does without labor, and is no more fitted than the balloon for rapid horizontal flight. If its engine stops, it must fall with deathly violence, for it can neither glide like the aeroplane, nor float like the balloon. The helicopter is much easier to design than the aeroplane, but is worthless when done. . .

—WILBUR WRIGHT
1909

In the opinion of competent experts it is idle to look for a commercial future for the flying machine. There is, and always will be, a limit to its carrying capacity. . . . Some will argue that because a machine will carry two people, another may be constructed that will carry a dozen, but those who make this contention do not understand the theory.

—W. J. Jackman and Thomas Russell
*Flying Machines: Construction and Operation*
1910

I would attack any squadron blockading a port. Nothing could prevent me from dropping out of the clear blue sky on to a battleship with 400 kilos of explosives in the cockpit. Of course it is true that the pilot would be killed, but everything would blow up, and that's what counts.

—Jules Vedrines
pre-1914

A new impetus was given to aviation by the relatively enormous power for weight of the atomic engine; it was at last possible to add Redmaynes's ingenious helicopter ascent and descent engine to the vertical propeller that had hitherto been the sole driving force of the aeroplane without over-weighting the machine, and men found themselves possessed of an instrument of flight that could hover or ascend or descend vertically and gently as rush wildly through the air. The last dread of flying vanished.

As the journalists of the time phrased it, this was the epoch of the Leap into the Air. The new atomic aeroplane became indeed a mania; everyone of means was frantic to possess a thing so controllable, so secure and so free from the dust and danger of the road, and in France in the year 1943 thirty thousand of these new aeroplanes were manufactured and licensed, and soared humming softly into the sky.

—H. G. Wells
*The World Set Free*
1914

The aeroplane is an invention of the devil and will never play any part in such a serious business as the defence of the nation, my boy.

—SIR SAM HUGHES
**Canadian Minister of Militia and Defence**
*to J.A.D. McCurdy, who had approached the minister with the idea of starting an air service*
**August 1914**

As a method of sending a missile to the higher, and even to the highest parts of the earth's atmospheric envelope, Professor Goddard's rocket is a practicable and therefore promising device. It is when one considers the multiple-charge rocket as a traveler to the moon that one begins to doubt . . . for after the rocket quits our air and really starts on its journey, its flight would be neither accelerated nor maintained by the explosion of the charges it then might have left. Professor Goddard, with his "chair" in Clark College and countenancing of the Smithsonian Institution, does not know the relation of action to re-action, and of the need to have something better than a vacuum against which to react . . . Of course he only seems to lack the knowledge ladled out daily in high schools.

—THE *NEW YORK TIMES* EDITORIAL
**13 January 1920**

Further investigation and experimentation have confirmed the findings of Isaac Newton in the 17th century, and it is now definitely established that a rocket can function in a vacuum as well as in an atmosphere. The Times regrets the error.

—THE *NEW YORK TIMES*
*regards their 1920 editorial*
**17 July 1969**

Within the next few decades, autos will have folding wings that can be spread when on a straight stretch of road so that the machine can take to the air.

—EDWARD "EDDIE" RICKENBACKER
***Popular Science***
**July 1924**

The aeroplane is tragically unsuited for ocean service.

—Dr Hugo Eckener
*dirigible advocate*
**1926**

Even though the release was pulled, the rocket did not rise at first, but the flame came out, and there was a steady roar. After a number of seconds it rose, slowly until in cleared the frame, and then at express-train speed, curving over to the left, and striking the ice and snow, still going at a rapid rate. It looked almost magical as it rose, without any appreciably greater noise or flame, as if it said, "I've been here long enough; I think I'll be going somewhere else, if you don't mind."

—Robert Goddard
*regards the first rocket flight using liquid propellants at Aunt Effie's farm*
**17 March 1926**

It is my contention that an agent ideal to the use of the scientific militarist, for both the air raid and the long distance bombardment is now in the process of development; that its eventual perfection is but a matter of time; and its use in warfare is certain to occur. I refer to the rocket. The perfection of the rocket in my opinion will give to future warfare the horror unknown in previous conflicts and will make possible destruction of nations, in a cool, passionless and scientific fashion.

—David Lasser
**22 October 1931**

There is no hope for the fanciful idea of reaching the moon because of insurmountable barriers to escaping the earth's gravity.

—Dr. F. R. Moulton
**University of Chicago astronomer**
**1932**

There is not in sight any source of energy that would be a fair start toward that which would be necessary to get us beyond the gravitative control of the earth.

—Forest Ray Moulton
**astronomer**
**1935**

If there is a possibility of cosmonautics, Man will not hesitate to leave the Earth to launch himself into interplanetary space at the risk of losing his own life.

—G. A. Crocco
*at the Fifth Volta Conference*
1935

The whole procedure [of shooting rockets into space] . . . presents difficulties of so fundamental a nature, that we are forced to dismiss the notion as essentially impracticable, in spite of the author's insistent appeal to put aside prejudice and to recollect the supposed impossibility of heavier-than-air flight before it was actually accomplished.

—Sir Richard van der Riet Wooley
British astronomer
reviewing P.E. Cleator's "Rockets in Space," *Nature*
March 14, 1936

The Americans cannot build aeroplanes. They are very good at refrigerators and razor blades.

—Hermann Goering
German Air Force Minister report to Hitler
1940

Mark my word: A combination airplane and motorcar is coming. You may smile. But it will come . . .

—Henry Ford
Chairman, Ford Motor Company
1940

It's only the beginning but the implications are terrific.

—Gerald Sayer
*first flight in the Gloster-Whittle E28 jet*
1941

Automobiles will start to decline almost as soon as the last shot is fired in World War II. The name of Igor Sikorsky will be as well known as Henry Ford's, for his helicopter will all but replace the horseless carriage as the new means of popular transportation. Instead of a car in every garage, there will be a helicopter. . .. These "copters" will be so safe and will cost so little to produce that small models will be made for teenage youngsters. These tiny 'copters, when school lets out, will fill the sky as the bicycles of our youth filled the prewar roads.

—Harry Bruno
**aviation publicist**
**1943**

Gliders. . . [will be] the freight trains of the air. . .. We can visualize a locomotive plane leaving LaGuardia Field towing a train of six gliders in the very near future. By having the load thus divided it would be practical to unhitch the glider that must come down in Philadelphia as the train flies over that place—similarly unhitching the loaded gliders for Washington, for Richmond, for Charleston, for Jacksonville, as each city is passed—and finally the air locomotive itself lands in Miami. During that process it has not had to make any intermediate landings, so that it has not had to slow down.

—Grover Loening
**1944**

The most fascinating aspect of successfully launching a satellite would be the pulse quickening stimulation it would give to considerations of interplanetary travel. Whose imagination is not fired by the possibility of voyaging out beyond the limits of our earth, traveling to the Moon, to Venus and Mars? Such thoughts when put on paper now seem like idle fancy. But, a man-made satellite, circling our globe beyond the limit of the atmosphere is the first step. The other necessary steps would surely follow in rapid succession. Who would be so bold as to say that this might not come within our time?

—Louis N. Ridenour
**"RAND's Role in the Evolution of Balloon and Satellite Observation Systems and Related U.S. Space Technology"**
**February 1947**

If you are in trouble anywhere in the world, an airplane can fly over and drop flowers, but a helicopter can land and save your life.

—Igor Sikorsky
**1947**

Satellite vehicles represent a rather fearsome foresight of future wars of nerves, in which aggressive nations could put their pilotless missiles into frictionless satellite motion round the earth for all to see and fear, with the constant threat of guiding them down to a target.

—W. F. Hilton
*High-Speed Aerodynamics*
1952

If we were to start today on an organized and well-supported space program I believe a practical passenger rocket can be built and tested within ten years.

—Dr. Wernher von Braun
on the "Tomorrowland" segment of TV show *Disneyland*
9 March 1955

Space travel is utter bilge.

—Sir Richard van der Riet Wooley
*on assuming the post of Astronomer Royal of Britain*
1956

Space travel is bunk.

—Sir Harold Spencer Jones
Astronomer Royal of Britain
*two weeks before the launch of Sputnik*
1957

We have all the prerequisities to build an aircraft powered by an atomic engine, in the near future.

—E. P. Slavsky
Chief of the Soviet atomic energy effort
reported in *Flying* magazine at the start of the article "Nuclear Power for Aircraft: Though many problems still exist, the A-powered plane's future looks brighter"
June 1957

Supersonic airplanes have carried men at more than 2,000 miles per hour and there are reasons to believe that this speed will be doubled by 1960 or so.

—Igor Sikorsky
14 January 1958

Men who have worked together to reach the stars are not likely to descend together into the depths of war and desolation.

—LYNDON BAINES JOHNSON
Then U.S. Senator
*addressing the U.N. General Assembly*
1958

[Before man reaches the moon] your mail will be delivered within hours from New York to California, to England, to India or to Australia by guided missiles. . . . We stand on the threshold of rocket mail.

—ARTHUR E. SUMMERFIELD
AP wire report
23 January 1959

The first man-made satellite to orbit the earth was named Sputnik. The first living creature in space was Laika. The first rocket to the moon carried a red flag. The first photograph of the far side of the moon was made with a Soviet camera. If a man orbits the earth this year his name will be Ivan

—JOHN F. KENNEDY
Then U.S. Senator
in an issue of *Missiles and Rockets*
(actually written by Edward O. Welsh)
1960

Many years ago the British explorer George Mallory, who was to die on Mount Everest, was asked why did he want to climb it. He said "Because it is there." Well, space is there, and we're going to climb it, and the moon and the planets are there, and new hopes for knowledge and peace are there.

—PRESIDENT JOHN F. KENNEDY
*address on the nation's space effort, Rice University, Houston, Texas*
September 12, 1962

The Post-Apollo manned space flight program is focusing on a 100-man Earth-orbiting station with a multiplicity of capabilities varying from development of earth resources to astronomy. . . . The schedule under consideration contemplates a launch of the first module of the large space station, with perhaps as many as 12 men, by 1975. Using the concept of modularity, NASA's advanced manned mission planners for see the gradual, incremental buildup of the initial station to a large base accommodating 100 men by 1980.

—*AVIATION WEEK & SPACE TECHNOLOGY*
**24 February 1969**

By the year 2000 we will undoubtedly have a sizable operation on the Moon, we will have achieved a manned Mars landing and it's entirely possible we will have flown with men to the outer planets.

—DR. WERNHER VON BRAUN
**1969**

If an elderly but distinguished scientist says that something is possible he is almost certainly right, but if he says that it is impossible he is very probably wrong.

—ARTHUR C. CLARKE
**in the *New Yorker* magazine**
**9 August 1969**

In all the history of mankind there will be only one generation which will be the first to explore the solar system, one generation for which, in childhood the planets are distant and indistinct discs moving through the night, and for which in old age the planets are places, diverse new worlds in the course of exploration. There will be a time in our future history when the solar system will be explored and inhabited by men who will be looking outward toward the first trip to the stars. To them and to all who come after us, the present moment will be a pivotal instant in the history of mankind.

—CARL SAGAN
*London lecture at Eugene, Oregon*
1970

Future growth potential looks unlimited . . . one gross weight doubling, and possibly two, is predicted by 1985; nuclear power can drive [the C-5A's] optimum weight to 5 or 10 million pounds before the year 2000.

—F. A. CLEVELAND
1970

It's too bad, but the way American people are, now that they have all this capability, instead of taking advantage of it, they'll probably just piss it all away.

—PRESIDENT LYNDON BAINES JOHNSON
*regards the end of the Apollo program*

Before another century is done it will be hard for people to imagine a time when humanity was confined to one world, and it will seem to them incredible that there was ever anybody who doubted the value of space and wanted to turn his or her back on the Universe.

—ISAAC ASIMOV
1979

It is marvelous indeed to watch on television the rings of Saturn close; and to speculate on what we may yet find at galaxy's edge. But in the process, we have lost the human element; not to mention the high hope of those quaint days when flight would create "one world." Instead of one world, we have "star wars," and a future in which dumb dented human toys will drift mindlessly about the cosmos long after our small planet's dead.

—GORE VIDAL
*On Flying*
1987

Our goal: To place Americans on Mars and to do it within the working lifetimes of scientists and engineers who will be recruited for the effort today. And just as Jefferson sent Lewis and Clark to open the continent, our commitment to the Moon/Mars initiative will open the Universe. It's the opportunity of a lifetime, and offers a lifetime of opportunity.

—PRESIDENT GEORGE BUSH
2 February 1990

All civilizations become either spacefaring or extinct.

—PROFESSOR CARL SAGAN
*Pale Blue Dot: A Vision of the Human Future in Space*
1994

It is the last day, barring unforeseen circumstances, that we will not have a human presence in space.

—RICHARD LABRODE
*U.S. flight director for the International Space Station, at mission control outside Moscow*
31 October 2000

But the fact that some geniuses were laughed at does not imply that all who are laughed at are geniuses. They laughed at Columbus, they laughed at Fulton, they laughed at the Wright Brothers. But they also laughed at Bozo the Clown.

—PROFESSOR CARL SAGAN

# Women Fly

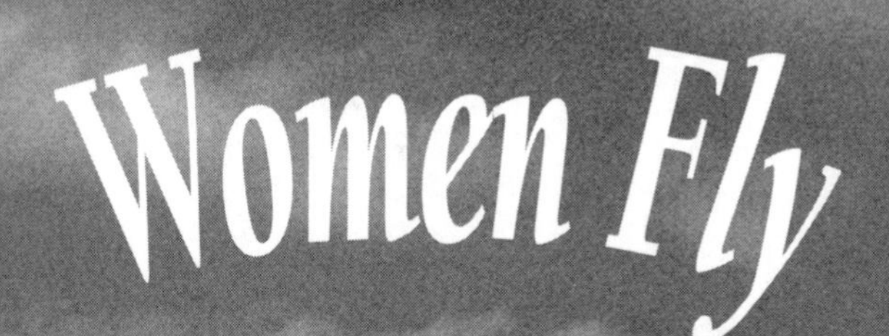

The air is the only place free from prejudices.

—Bessie Coleman
*who had to go to France to learn how to fly when Americans would not instruct a black lady*
1921

Flying does not rely so much on strength, as on physical and mental coordination.

—Elise Deroche
*first woman to solo an airplane.*

Flying is the best possible thing for women.

—Elise Deroche

There is a world-old controversy that crops up whenever women attempt to enter a new field. Is a woman fit for that work? It would seem that a woman's success in any particular field would prove her fitness for that work, without regard to theories to the contrary.

—Ruth Law
1920

The aeroplane should open a fruitful occupation for women. I see no reason they cannot realize handsome incomes by carrying passengers between adjacent towns, from parcel delivery, taking photographs or conducting schools of flying.

—Harriet Quimby
June 1912

Aviation, this young modern giant, exemplifies the possible relationships of women with the creations of science.

—Amelia Earhart,
Purdue University, Lafayette, Indiana
1935

I have found that women are not only just as much interested as men are in flying, but apparently have less fear than the men have. At least, more women than men asked to go up with me. And when I took them up, they seemed to enjoy it.

—KATHERINE STINSON
quoted in *Katherine Stinson: The Flying Schoolgirl*
Debra L. Winegarten

This is not a time when women should be patient. We are in a war and we need to fight it with all our ability and every weapon possible. Women pilots, in this particular case, are a weapon waiting to be used.

—ELEANOR ROOSEVELT
1942

We realized what a spot we were in. We had to deliver the goods, or else there wouldn't ever be another chance for women pilots in any part of the service.

—CORNELIA FORT, WAFS,
1942

I can cure your men of walking off the [flight] program. Let's put on the girls.

—JACQUELINE COCHRAN

Any girl who has flown at all grows used to the prejudice of most men pilots who will trot out any number of reasons why women can't possibly be good pilots. . . . The only way to show the disbelievers, the snickering hangar pilots is to show them.

—CORNELIA FORT

There is a decided prejudice on the part of the general public against being piloted by a woman, and as great an aversion, partially because of this, by executives of those companies whose activities require employing pilots.

—LOUISE THADEN
Co-founder of the Ninety-Nines

If you have flown, perhaps you can understand the love a pilot develops for flight. It is much the same emotion a man feels for a woman, or a wife for her husband.

—LOUISE THADEN

Too often little attention is paid to individual talent. Instead, education goes on dividing people according to their sex, and putting them in little feminine or masculine pigeonholes . . . Girls are shielded and sometimes helped so much that they lose initiative and begin to believe the signs "Girls don't" and "Girls can't" which mark their paths. . . Consequently, it seems almost necessary to evolve different methods of instruction for them when they later take up the same subjects. For example, those courses which involve mechanical work may have to be explained somewhat differently to girls not because girls are inherently not mechanical, but because normally they have learned little about such things in the course of their education.

—AMELIA EARHART
*The Fun of It*
1932

To a psychoanalyst, a woman pilot, particularly a married one with children, must prove an interesting as well as an inexhaustible subject. Torn between two loves, emotionally confused, the desire to fly an incurable disease eating out your life in the slow torture of frustration—she cannot be a simple, natural personality.

—LOUISE THADEN

The men flyers have given out the impression that aeroplaning is very perilous work, something that an ordinary mortal should not dream of attempting. But when I saw how easily the men flyers manipulated their machines I said I could fly.

—HARRIET QUIMBY

If you can't see any opportunities where you are now, don't waste your time criticizing the darkness. . . . Light a candle to find your way out.

—ARLENE FELDMAN
Regional Director FAA

Aviation is still considered a man's world by many. The time to reach young ladies is during their first years of school. Research has shown that although children may change their minds several times about their eventual careers, the possibilities of them selecting a non-traditional role must be nurtured at an early age.

—DR. PEGGY BATY
**Founder of Women In Aviation International**

In the early days they said I was trying to make a statement, but I was just trying to make a living.

—CAPTAIN BONNIE TIBURZI, AMERICAN AIRLINES
*first woman hired by a major airline*

I was annoyed from the start by the attitude of doubt by the spectators that I would never really make the flight. This attitude made me more determined than ever to succeed.

—HARRIET QUIMBY
*just prior to her flight across the English Channel*
**1912**

It is not easy to be the best. You must have the courage to bear pain, disappointment, and heartbreak. You must learn how to face danger and understand fear, yet not be afraid. You establish your goal, and no matter what deters you along the way, in your every waking moment you must say to yourself, "I could do it."

—BETTY SKELTON
*first lady of aerobatics*

Have confidence in yourself and tell yourself "you can" twice for every time you are told "you can't." Confidence that you can succeed is everything. Take every negative remark as a challenge to achieve more and progress to newer heights. You are able to do anything you believe you can do. You might even surprise yourself.

—ALINDA WIKERT
*first female owner and CEO of an airline*

Flying is not about whether the pilot is a man or a woman. It is about the results of the actions imposed by the pilot and the responses returned by the aircraft. The aircraft does not know or understand gender. It only knows the difference in a true pilot, and one who was perhaps not meant to fly.

—CAPTAIN JENNIFER KAYE, AIR NATIONAL GUARD
**2000**

I wanted to be a hairdresser when I grew up. I'd sit on the back of the sofa and my mother would sit in front of me, and as long as it didn't involve scissors or dye, she'd let me do whatever I wanted to her hair. All the women in my life were nurses, hairdressers, or secretaries, and that's why I thought my father would not support me in being a pilot. I can remember asking him, "what would you think if I told you I wanted to be a pilot when I grew up?" expecting him to say no or disagree. He said, "I think that would be fantastic." Had he not said those words, I don't know what would have happened to me.

—SUSAN STILL, LIEUTENANT, UNITED STATES NAVY
**Combat Pilot and Astronaut**
**quoted in *Women and Flight***
**Carolyn Russo**
**1997**

Space is for everybody. It's not just for a few people in science or math, or for a select group of astronauts. That's our new frontier out there, and it's everybody's business to know about space.

—CHRISTA MCAULIFFE
**6 December 1985**

Because of [Amelia Earhart], we had more women available to fly in the 1940s to help us get through World War II. And because of these women, women of my generation are able to look back and say, "Hey, they did it. They even flew military airplanes, we can do it, too."

—AIR FORCE COL. EILEEN COLLINS
**television interview "100 Years of Great Women"**
**on ABC with Barbara Walters**
**30 April 1999**

I'm honored to be the first woman to have the opportunity to command the shuttle. I don't really think about that on a day-to-day basis because I really don't need to.

—AIR FORCE COL. EILEEN COLLINS
*first female Space Shuttle commander*
**24 June 1999**

There was my mom and I had a wife for a long time and now there is my fiancée. Eileen is in a long line of women who have given me orders.

—JEFFREY S. ASHBY
**shuttle pilot**
*regards flying under Eileen Collins' command*

My daughter just thinks that all moms fly the space shuttle.

—AIR FORCE COL. EILEEN COLLINS
**1999**

# Combat

There's something wonderfully exciting about the quiet sing song of an aeroplane overhead with all the guns in creation lighting out at it, and searchlights feeling their way across the sky like antennae, and the earth shaking snort of the bombs and the whimper of shrapnel pieces when they come down to patter on the roof.

—John Dos Passos
letter written in Bossano, Italy, while serving in the American Red Cross Ambulance Service to his friend Marvin T. Rumsey
18 February 1918

Eyesight and seeing the enemy first, or at least in time to take correct tactical maneuvers was very important. However, most important is the guts to plough through an enemy or enemies, and fight it out. There are no foxholes to hide in . . . there is no surrendering. I know of no Navy fighter pilot in the war who turned tail and ran. If one did, he would lose his wings and be booted out of the service for cowardice.

—Richard H. May, USN

It's just like being in a knife fight in a dirt-floor bar. If you want to fix a fella, the best way to do it is to get behind him and stick him in the back. It's the same in an air fight. If you want to kill that guy, the best thing to do is get around behind him where he can't see you . . . and shoot him.

—Captain William O'Brian
357th Fighter Group, USAAF

I think that the most important features of a fighter pilot are aggressiveness and professionalism. They are both needed to achieve the fighter pilot's goal: the highest score within the shortest time, with the least risk to himself and his wingman.

—Colonel Gidi Livni
Israeli Air Force

The most important thing to a fighter pilot is speed; the faster an aircraft is moving when he spots an enemy aircraft, the sooner he will be able to take the bounce and get to the Hun. If you have any advantage on him, keep it and use it. When attacking, plan to overshoot him if possible, hold fire until within range, then shoot and clobber him down to the last instant before breaking away. It's like sneaking up behind someone and hitting them with a baseball bat.

—Duane W. Beeson
**P-51 pilot, 4th Fighter Group**

Once committed to an attack, fly in at full speed. After scoring crippling or disabling hits, I would clear myself and then repeat the process. I never pursued the enemy once they had eluded me. Better to break off and set up again for a new assault. I always began my attacks from full strength, if possible, my ideal flying height being 22,000 ft because at that altitude I could best utilize the performance of my aircraft. Combat flying is based on the slashing attack and rough maneuvering. In combat flying, fancy precision aerobatic work is really not of much use. Instead, it is the rough maneuver which succeeds.

—Colonel Erich "Bubi" Hartmann, GAF
*aka Karaya One, world's leading ace, with 352 victories in W.W. II*

To be a good fighter pilot, there is one prime requisite—think fast, and act faster.

—Major John T. Godfrey, USAAF

I had no system of shooting as such. It is definitely more in the feeling side of things that these skills develop. I was at the front five and a half years, and you just got a feeling for the right amount of lead.

—Lt. General Guenther Rall, GAF

I am not a good shot. Few of us are. To make up for this I hold my fire until I have a shot of less than 20 degrees deflection and until I'm within 300 yards. Good discipline on this score can make up for a great deal.

—Lt. Colonel John C. Meyer, USAAF

Mark Twain said, "Courage is the mastery of fear, resistance to fear, not the absence of fear." At times the nearness of death brings an inexplicable exhilaration which starts the adrenaline flowing and results in instant action. The plane becomes an integral part of the pilot's body, it is strapped to his butt, and they become a single fighting machine.

—R. M. LITTLEFIELD
*Double Nickel-Double Trouble*
1993

What you have at that moment is—well, it may sound strange, but it's actually fun. The other guy has his chance, too, and you've got to get him before he gets you. Yes, I think it is the most exciting fun in the world.

—LT. COL. ROBERT B. "WESTY" WESTBROOK, USAAF
*one of the leading aces of the Pacific*
*Los Angeles Examiner*
20 June 1944

And I have yet to find one single individual who has attained conspicuous success in bringing down enemy aeroplanes who can be said to be spoiled either by his successes or by the generous congratulations of his comrades. If he were capable of being spoiled he would not have had the character to have won continuous victories, for the smallest amount of vanity is fatal in aeroplane fighting. Self-distrust rather is the quality to which many a pilot owes his protracted existence.

—CAPTAIN EDWARD V. "EDDIE" RICKENBACKER, USAS
*Fighting the Flying Circus*
1919

My habit of attacking Huns dangling from their parachutes led to many arguments in the mess. Some officers, of the Eton and Sandhurst type, thought it was "unsportsmanlike" to do it. Never having been to a public school, I was unhampered by such considerations of form. I just pointed out that there was a bloody war on, and that I intended to avenge my pals.

—IRA "TAFFY" JONES, RFC
W.W. I

The greater issues were beyond us. We sat in a tiny cockpit, throttle lever in one hand, stick in the other. At the end of our right thumb was the firing button, and in each wing were four guns. We aimed through an optical gunsight, a red bead in the middle of a red ring. Our one concern was to boot out the enemy.

—Group Captain Peter Townsend, RAF

My pilot pointed to his left front and above, and looking in the direction he pointed, I saw a long dark brown form fairly streaking across the sky. We could see that it was a German machine, and when it got above and behind our middle machine, it dived on it for all the world like a huge hawk on a hapless sparrow.

—James McCudden, VC, RFC

The experienced fighting pilot does not take unnecessary risks. His business is to shoot down enemy planes, not to get shot down. His trained hand and eye and judgment are as much a part of his armament as his machine-gun, and a fifty-fifty chance is the worst he will take—or should take—except where the show is of the kind that . . . justifies the sacrifice of plane or pilot.

—Captain Edward "Eddie" Rickenbacker

I saw the lightnings gleaming rod.
Reach forth and write upon the sky
The awful autograph of God.

—Joaquin Miller
*The Ship in the Desert*

There was only one catch and that was Catch22, which specified that a concern for one's safety in the face of dangers that were real and immediate was the process of a rational mind. Orr was crazy and could be grounded. All he had to do was ask, and as soon as he did, he would no longer be crazy and would have to fly more missions. Orr would be crazy to fly more missions and sane if he didn't, but if he was sane he had to fly them. If he flew them he was crazy and didn't have to; but if he didn't want to he was sane and had to.

—Joseph Heller
*Catch22*
1955

He who has the height controls the battle.
He who has the sun achieves surprise.
He who gets in close shoots them down.

—ANON

I belong to a group of men who fly alone. There is only one seat in the cockpit of a fighter airplane. There is no space alotted for another pilot to tune the radios in the weather or make the calls to air traffic control centers or to help with the emergency procedures or to call off the airspeed down final approach. There is no one else to break the solitude of a long cross-country flight. There is no one else to make decisions.

I do everything myself, from engine start to engine shutdown. In a war, I will face alone the missiles and the flak and the small-arms fire over the front lines.

If I die, I will die alone.

—RICHARD BACH
*Stranger to the Ground*
1963

The ordinary air fighter is an extraordinary man and the extraordinary air fighter stands as one in a million among his fellows.

—THEODORE ROOSEVELT

Most healthy young men or women from sixteen to forty years of age can be taught to fly an ordinary airplane. A great majority of these may become very good pilots for transport- or passenger-carrying machines in time of peace; but the requirements for a military aviator call for more concentrated physical and mental ability in the individual than has ever been necessary in any calling heretofore.

—BRIGADIER GENERAL WILLIAM 'BILLY' MITCHELL
*Skyways*
1930

For most of the time carrier aviation is more challenging than flying in a spacecraft.

—JAMES LOVELL
Astronaut

Night flying in blacked-out Britain is like flying up a cow's ass.

—SQUADRON LEADER EARL BRACKEN, RAF

We train young men to drop fire on people. But their commanders won't allow them to write "fuck" on their airplanes because? It's obscene!

—COLONEL WALTER E. KURTZ
in the 1979 movie *Apocalypse Now*

Yea though I fly through the valley of the shadow of death. . . I fear no evil . . . for I fly the biggest, baddest, meanest, fastest motherfucker in the whole damn valley.

—ANON

The fortress inspired a tremendous confidence. It was the only propeller driven aircraft I have flown that was completely viceless; there were no undesirable flight characteristics. The directional stability was excellent and, properly trimmed, the B-17 could be taken off, landed and banked without change of trim.

—LT. JAMES W. JOHNSON, USAAF

I'm waiting to be told how cobras, hooks, or vectored thrust help in combat. They're great at air shows, but zero energy is a fighter pilot's nightmare. Shoot your opponent down and his number two will be on your tail thinking it's his birthday—a target hanging there in the sky with zero energy.

—NED FIRTH
Eurofighter

Being a stealth pilot is one of the most labor intensive and time constrained types of flying that I know. We have very strict timing constraints: to be where you are supposed to be all the time, exactly on time, and that has to be monitored by the pilot. For example, during a bomb competition in training in the US, I dropped a weapon that landed 0.02 seconds from the desired time, and finished third!

—LT. COL. MILES POUND, USAF

Do unto the other feller the way he'd like to do unto you, an' do it fust [sic].

—E. N. Westcott
*David Harum*

When I took over my wing [in Vietnam], the big talk wasn't about the MIGs, but about the SAMs . . . I'd seen enemy planes before, but those damn SAMs were something else. When I saw my first one, there were a few seconds of sheer panic, because that's a most impressive sight to see that thing coming at you. You feel like a fish about to be harpooned. There's something terribly personal about the SAM; it means to kill you and I'll tell you right now, it rearranges your priorities . . . We had been told to keep our eyes on them and not to take any evasive move too soon, because they were heat-seeking and they, too would correct, so I waited until it was almost on me and then I rolled to the right and it went on by. It was awe inspiring . . . The truth is you never do get used to the SAMs; I had about two hundred fifty shot at me and the last one was as inspiring as the first. Sure I got cagey, and I was able to wait longer and longer, but I never got overconfident. I mean, if you're one or two seconds too slow, you've had the schnitzel.

—General Robin Olds, USAF

It is as though horror has frozen the blood in my veins, paralyzed my arms, and torn all thought from my brain with the swipe of a paw. I sit there, flying on, and continue to stare, as though mesmerised, at the Cauldron on my left.

—Ernst Udet
*My Life As Aviator*
1935

How this can happen is a mystery to us.

—Lieutenant-General Ray Henault
Canadian Chief of Defence staff
*regards the friendly fire deaths of four Canadian soldiers by a USAF F-16 in Afghanistan*
18 April 2002

*DICTA BOELCKE*

Try to secure advantages before attacking. If possible, keep the sun behind you.

Always carry through an attack when you have started it.

Fire only at close range, and only when your opponent is properly in your sights.

Always keep your eye on your opponent, and never let yourself be deceived by ruses.

In any form of attack it is essential to assail your opponent from behind.

If your opponent dives on you, do not try to evade his onslaught, but fly to meet it.

When over the enemy's lines never forget your own line of retreat.

For the Staffel: attack on principle in groups of four or six. When the fight breaks up into a series of single combats, take care that several do not go for one opponent.

—HAUPTMANN OSWALD BOELCKE
*Germany's first ace, who died in 1916 with 40 victories*
**1916**

Won't it be nice when all this beastly killing is over, and we can enjoy ourselves and not hurt anyone? I hate this game.

—CAPTAIN ALBERT BALL, RFC
*In letters to his father and fiancée. Ball was the first British ace idolized by the public, 44 victories when killed in action.*
**6 May 1917**

Whatever Boelcke told us was taken as Gospel!

—BARON MANFRED VON RICHTHOFEN

I will be like Boelcke.

—GERMAN PILOTS' MOTTO

He was neither seen nor heard as he fell, his body and his machine were never found. Where has he gone? By what wings did he manage to glide into immortality? Nobody knows: nothing is known. He ascended and never came back, that is all. Perhaps our descendents will say: He flew so high that he could not come down again.

—*L'ILLUSTRATION* NEWSPAPER
*Obituary of Capitaine Georges Marie Ludovic Jules Guynemer, 53 victories W.W. I*
**6 October 1917**

You must take the war to the enemy. You must attack and go on attacking all the time.

—MAJOR WILLY OMER FRANÇOIS JEAN COPPENS DE HOUTHULST, BELGIAN AIR SERVICE
*37 victories W.W. I*

There is no question about the hereafter of men who give themselves in such a cause. If I am called upon to make it, I shall go with a grin of satisfaction and a smile.

—LIEUTENANT DAVID ENDICOTT PUTNAM
*America's first "Ace of Aces," in a letter to his mother. He was shot down by German ace Georg von Hantelmann.*
**12 September 1918**

It was war. We were defending our country. We had a strict code of honor: you didn't shoot down a cripple and you kept it a fair fight.

—CAPTAIN WILFRID REID "WOP" MAY, RFC
*13 victories W.W. I*

I put my bullets into the target as if I placed them there by hand.

—CAPITAINE RENÉ PAUL FONCK, FRENCH AIR SERVICE
*75 victories W.W. I*

It is wonderful how cheered a pilot becomes after he shoots down his first machine; his morale increases by at least 100 percent.

—CAPTAIN JAMES IRA THOMAS "TAFFY" JONES, RFC
*37 victories in 3 months W.W. I*

# Air Power

Thank God men cannot as yet fly and lay waste the sky as well as the earth!

—HENRY DAVID THOREAU

If our airforces are never used, they have achieved their finest goal.

—GENERAL NATHAN F. TWINING, USAF

The greatest contributor to the feeling of tension and fear of war arose from the power of the bombing aeroplane. If all nations would consent to abolish air bombardment . . . that would mean the greatest possible release from fear.

—ERNEST RUTHERFORD

The development of air power in its broadest sense, and including the development of all means of combating missiles that travel through the air, whether fired or dropped, is the first essential to our survival in war.

—VISCOUNT HUGH M. TRENCHARD
1946

You can shoot down every MiG the Soviets employ, but if you return to base and the lead Soviet tank commander is eating breakfast in your snack bar, Jack, you've lost the war.

—ANONYMOUS A-10 PILOT

No enemy bomber can reach the Ruhr. If one reaches the Ruhr, my name is not Goering. You may call me Meyer.

—HERMANN GOERING, GERMAN AIR FORCE MINISTER
1939

. . . when I look round to see how we can win the war I see that there is only one sure path . . . and that is absolutely devastating, exterminating attack by very heavy bombers from this country upon the Nazi homeland. We must be able to overwhelm them by this means, without which I do not see a way through.

—BRITISH PRIME MINISTER WINSTON CHURCHILL
in a letter to Minister of Aircraft Production Lord Beaverbrook
July 1940

War is a nasty, dirty, rotten business. It's all right for the Navy to blockade a city, to starve the inhabitants to death. But there is something wrong, not nice, about bombing that city.

—SIR ARTHUR 'BOMBER' HARRIS
**Marshal of the Royal Air Force**

Are you aware it is private property? Why you'll be asking me to bomb Essen next.

—SIR KINGSLEY WOOD
**British Secretary of State for Air**
*regards plans to set fire to the Black Forest*
**September 1939**

Their element is to attack, to track, to hunt, and to destroy the enemy. Only in this way can the eager and skillful fighter pilot display his ability. Tie him to a narrow and confined task, rob him of his initiative, and you take away from him the best and most valuable qualities he posesses: aggressive spirit, joy of action, and the passion of the hunter.

—GENERAL ADOLF GALLAND
**Luftwaffe**

Adolf Galland said that the day we took our fighters off the bombers and put them against the German fighters, that is, went from defensive to offensive, Germany lost the air war. I made that decision and it was my most important decision during World War II. As you can imagine, the bomber crews were upset. The fighter pilots were ecstatic.

—GENERAL JAMES H. DOOLITTLE

Never abandon the possibility of attack. Attack even from a position of inferiority, to disrupt the enemy's plans. This often results in improving one's own position.

—GENERAL ADOLF GALLAND
**Luftwaffe**
**15 March 1941**

During the Battle of Britain the question "fighter or fighter-bomber?" had been decided once and for all: The fighter can only be used as a bomb carrier with lasting effect when sufficient air superiority has been won.

—General Adolf Galland,
Luftwaffe
*The First and the Last*
1954

The Navy can lose us the war, but only the Air Force can win it. The fighters are our salvation, but the bombers alone provide the means of victory.

—Prime Minister Winston Churchill

The air fleet of an enemy will never get within striking distance of our coast as long as our aircraft carriers are able to carry the preponderance of air power to sea.

—Rear Admiral W. A. Moffet
Chief of the US Bureau of Aeronautics
October 1922

I am the bomber 17 —
Proud machine—sleek and powerful,
Made by man to kill his foe,
Made of steel and wood and metal,
Built to fight and drop destruction . . .

—Robert Cromwell,
*Skyward: A Ballad of the Bomber*

I am purely evil;
Hear the thrum
of my evil engine;
Evilly I come.
The stars are thick as flowers
In the meadows of July;
A fine night for murder
Winging through the sky.

—Ethel Mannin
'Song of the Bomber'

Four other pieces of equipment that most senior officers came to regard as among the most vital to our success in Africa and Europe were the bulldozer, the jeep, the 2½-ton truck, and the C-47 airplane. Curiously, none of these is designed for combat.

—DWIGHT D. EISENHOWER

Give me fifty DC-3's and the Japanese can have the Burma Road.

—CHIANG KAI-SHEK

I wanted to soar through the air

—YUJI NISHIZAWA,
*after hijacking All Nippon Airways flight 61 and stabbing the captain to death in order to try and fly the B-747 himself*
**July 1999**

Either end your life while praying, seconds before your target, or make your last words: 'There is no God but God, Mohammad is His messenger.'

—TRANSLATED FROM WRITTEN INSTRUCTIONS FOR MOHAMED ATTA
*the alleged terrorist thought to be at the controls of American Airlines flight 11*

What do I tell the pilots to do?

—BARBARA OLSON
**CNN commentator**
*passenger on American Airlines flight 77, during a cell phone call to her husband, Justice Department official Theodore Olson*
**11 September 2001**

A group of us are going to do something.

—THOMAS E. BURNETT JR.
**Thoratec Corporation**
**Senior Vice President**
*and passenger on United Airlines Flight 93, cell phone call to his wife*
**11 September 2001**

We're going to rush the hijackers.

—JEREMY GLICK
**Software Executive**
*and passenger on United Airlines flight 93, last reported words from his cell phone call*
**11 September 2001**

Are you guys ready? Let's roll!

—TODD BEAMER
**Oracle software executive**
*and passenger on United Airlines flight 93, last reported words from his cell phone call*
**11 September 2001**

What's the sense of sending $2 million missiles to hit a $10 tent that's empty?

—PRESIDENT GEORGE W. BUSH
*Oval Office meeting*
**13 September 2001**

Is it likely that an aircraft carrier or a cruise missile is going to find a person?

—DONALD H. RUMSFELD,
**US Defense Secretary**
*regards questions on an air war to kill Osama bin Laden*
**23 September 2001**

# Bums on Seats

As of 1992, in fact—though the picture would have improved since then—the money that had been made since the dawn of aviation by all of this country's airline companies was zero. Absolutely zero.

—WARREN BUFFETT
**Billionaire investor**
**interview 1999**

People who invest in aviation are the biggest suckers in the world.

—DAVID G. NEELEMAN
*after raising a record $128 million to start New Air*
*(the then working name for what became jetBlue)*
**quoted in *Business Week***
**3 May 1999**

I don't believe in being the launch customer for anything.

—CARL MICHEL
**British Airways' commercial director**
*regards the then named Airbus A3XX (now A380)*
**February 2000**

In airplanes you have a choice between chocolate and vanilla. One year could be vanilla or it could be chocolate. I don't attach any relevance to which one.

—GORDON BETHUNE
**Chairman and CEO Continental Airlines**
*regards buying Boeing or Airbus products*
**2000**

The sun is now climbing from the west. In winter it is possible to leave London after sunset, on the evening Concorde for New York, and watch the sun rise out of the west. Flying at Mach 2 at these latitudes will cause the sun to set in the west at three times its normal rate, casting, as it does so, a vast curved shadow of the earth, up and ahead of the aircraft.

—FIRST OFFICER CHRISTOPHER ORLEBAR
**British Airways**
*regards the Concorde*

An aircraft which is used by wealthy people on their expense accounts, whose fares are subsidized by much poorer taxpayers.

—DENIS HEALEY
British Labour Party
*regards the Concorde*

Its operation in a world beset by fuel and energy crises makes no sense at all.

—SENATOR ALAN CRANSTON OF CALIFORNIA
*regards the Concorde*
1974

Without doubt, Concorde died yesterday at the age of 31. All that will remain is the myth of a beautiful white bird.

—LE FIGARO EDITORIAL
*the day after Air France 4590 crashed after takeoff from Charles de Gaulle aerodrome*
26 July 2000

The Boeing 747 is so big that it has been said that it does not fly; the earth merely drops out from under it.

—CAPTAIN NED WILSON
Pan American World Airlines

The Boeing 747 is the commuter train of the global village.

—H. TENNEKES
*The Simple Science of Flight*
1996

If anybody ever flied to the moon, the very next day Trippe will ask the Civil Aeronautics Board to authorize regular service.

—JAMES M. LANDIS
*regards Pan American World Airlines*

This is one hell of a good deal for United Airlines.

—RICHARD FERRIS
Chairman United Airlines
*after buying Pan Am's Far East route network*

People Express is clearly the archetypical deregulation success story and the most spectacular of my babies. It is the case that makes me the proudest.

—Alfred Kahn
**Professor of Political Economy, Cornell University**
***Time* magazine**
**13 January 1986**

It's a great day for TWA.

—William Compton
**President Trans World Airlines Inc.**
*on the day that U.S. District Judge Sue L. Robinson approved American Airlines' $200 million emergency financing plan, and cleared the way for an auction of America's longest-flying airline*
**27 January 2001**

It was the first airplane . . . that could make money just by hauling passengers.

—C. R. Smith
**President of American Airlines**
*regarding the DC-3. The DC-3 specifications were shaped by American Airlines*

You've got to treat people as equals, and make them feel like it's their company. I don't know if I've had any impact or helped persuade Frank [to sell Eastern]. But, I can tell you, there were many discussions on the subject.

—Michael Milken

If you would look up bad labor relations in the dictionary, you would have an American Airlines logo beside it.

—U.S. District Judge Joe Kendall
*issuing a restraining order against an American Airlines pilot union sickout*
**10 February 1999**

The greatest sin of airline management of the last 22 years is to say, "It's all labor's fault."

—Donald Carty
**Chairman and CEO American Airlines**
**12 August 2002**

Americans love rising-from-the-ashes stories. They love the underdog coming back. We're going to take a tarnished brand name and bring it back to a high degree of luster.

—Martin R. Shugrue Jr.
**President and CEO, Pan American World Airways**
**1996**

In my own view, it was not merely uncomfortable, it was intolerable. It might perhaps have been endurable for a two-hour flight but an eight-hour flight is a totally different matter.

—Judge Gareth Edwards QC
*regards a 29-inch seat pitch. The judge upheld a compensation award made to Brian Horan after he suffered deep-vein thrombosis on his journey Manchester, England, to the Canadian ski resort of Calgary. Chester County Court*
**17 April 2002**

If you want to travel on the airline system, you give up your privacy. If you want your privacy, don't fly. Flying is voluntary.

—Robert Crandall
**Former CEO American Airlines**
*regards airline security*
**April 2002**

The tolerance of the public is diminishing. We're spending time on the wrong people. It's nutty. There has to be a better way. Why are we stripsearching Aunt Molly from Iowa and letting on Richard Reid?

—Donald Carty
**Chairman and CEO American Airlines**
**April 2002**

Every time we hit an air pocket and the plane dropped about five hundred feet (leaving my stomach in my mouth) I vowed to give up sex, bacon, and air travel if I ever made it back to terra firma in one piece.

—Erica Jong
***Fear of Flying***
**1973**

I don't mind flying. I always pass out before the plane leaves the ground.

—NAOMI CAMPBELL
**supermodel**
**2000**

So long as the airlines preserve their magic quality—including, above all, their safety and reliability—they will be guaranteed a significant role in the workings of the world. Science will never digitalize an embrace. Electronics will never convey the wavering eye of a negotiating adversary. Fiber-optic cable can do many things, but it cannot transport hot sand, fast snow, or great ruins.

—THOMAS PETZINGER, JR.
*Hard Landing*
**1995**

Since 1978 the record pretty well shows that no start-up airline . . . has really been successful, so the odds of jetBlue having long-term success are remote. I'm not going to say it can't happen because stranger things have happened, but I personally believe P.T. Barnum was, in that respect, correct.

—GORDON BETHUNE
**CEO Continental Airlines**
*regards the 70% rise in jetBlue's stock price in the days after its IPO. Continental's annual shareholder meeting*
**17 April 2002**

He's a nut job, but a focused nut job.

—ROBERT LAND
**JetBlue government affairs director,**
*regards boss David Neeleman, quoted in Fortune magazine*
**May 2001**

I'm here to tell you that I am proud of a couple of things. First, I am very good at projectile vomiting. Second, I've never had a really serious venereal disease.

—HERB KELLEHER
*addressing the Wings Club in New York regards his time as CEO of Southwest Airlines*
**2001**

My wife was a stewardess, flying DC-3s. That's how we met. She knew what was going on. So when we got married, I made her a promise—the obvious one. And I've kept it.

—CAPTAIN ANSON HARRIS
**in the 1970 movie *Airport***

I did not fully understand the dread term "terminal illness" until I saw Heathrow for myself.

—DENNIS POTTER
**in *The Sunday Times***
**4 June 1978**

Anything that is white is sweet.
Anything that is brown is meat.
Anything that is grey, don't eat.

—STEPHEN SONDHEIM
*regards airline food*
**"Do I Hear a Waltz"**
**1965**

It's either expensive or it's crappy.

—JETBLUE AIRLINES SPOKESMAN
*regards airline food. Reported in the New York Times*
**26 June 2002**

The thing I miss about Air Force One is they don't lose my luggage.

—PRESIDENT GEORGE BUSH

# Humor

Though I Fly Through the Valley of Death I Shall Fear No Evil For I Am 80,000 feet and Climbing.

—SIGN OVER THE ENTRANCE TO THE SR-71 OPERATING LOCATION ON KADENA AFB, OKINAWA

Well boys, we've got three engines out, we've got more holes in us than a horse trader's mule, the radio is gone and we're leaking fuel and if we was flying any lower why we'd need sleigh bells on this thing . . . but we've got one thing on those Russkies. At this height why they might harpoon us but they dang sure ain't gonna spot us on no radar screen!

—MAJOR T. J. "KING" KONG
in the 1963 movie *Dr. Strangelove or: How I Learned to Stop Worrying and Love the Bomb*

Willie, how long can you tread water?

—COMMANDER RANDY 'DUKE' CUNNINGHAM, USN
*to his backseater after their F-4 took a missile hit over NVN and they dashed for the coast*

One of my life rules was to never give up a free ride when you're shark bait.

—LT. COL. DAN HOUSE
*SR-71 pilot describing his rescue from shark-infested waters by a native canoe*
*Lockheed SR-71: The Secret Missions Exposed*
1993

The Yo-Yo is very difficult to explain. It was first perfected by the well-known Chinese fighter pilot Yo-Yo Noritake. He also found it difficult to explain, being quite devoid of English.

—SQUADRON LEADER K. G. HOLLAND, RAF

Both optimists and pessimists contribute to the society. The optimist invents the aeroplane, the pessimist the parachute.

—GEORGE BERNARD SHAW

The weird thing is that I hate to fly, and the quote that I give people is that every time I get off a plane, I view it as a failed suicide attempt.

—BARRY SONNENFELD

*Tower:* Have a good trip.

*Pilot:* Make that a round trip . . .

—LLOYD LACE, USAAF
*Said before departing on C-46 missions, flying over "The Hump" (China – Burma – India)*
1944

Motor cut. Forced landing. Hit cow. Cow died. Scared me.

—DEAN SMITH
*telegraph to his chief,*
quoted by Amelia Earhart
*The Fun of It*
1932

Does anyone on board know how to fly a plane?

—ELAINE
speaking over the cabin speakers in the 1980 movie
*Airplane!*

If God had meant Icarus to fly, she would have given him a cloudy day.

—LEON M. WISE

If God had meant man to fly, He would never have given us the steam railway locomotive.

—A *SLIPPING THE SURLY BONDS: GREAT QUOTATIONS ON FLIGHT* READER'S LATE GREAT AUNT

If we love to fly so much, how come we're always in such a hurry to get there?

—LOUIE MANYAK

I hate to wake up and find my co-pilot asleep.

—MICHAEL TREACY

The most dangerous thing about flying is the risk of starving to death.

—DICK DEPEW

There are only two emotions in a plane: boredom and terror.

—ORSON WELLES

I know, but this guy doing the flying has no airline experience at all. He's a menace to himself and everything else in the air. . . . Yes, birds too.

—AIR TRAFFIC CONTROLLER IN THE 1980 MOVIE *AIRPLANE!*

Ted: "We're gonna have to come in pretty low on this approach."

Elaine: "Is that difficult?"

Ted: "Well sure it's difficult. It's part of every textbook approach. It's just something you have to do . . . when you land."

—FROM THE 1982 MOVIE *AIRPLANE II, THE SEQUEL*

When the art of radio communication between pilots and ATC is improved, the result will be vastly increased areas of significant misunderstandings.

—ROBERT LIVINGSTON
*Flying the Aeronca*
1981

Buttons . . . check. Dials . . . check. Switches . . . check. Little colored lights . . . check.

—THE BILL WATERSON COMIC CHARACTER CALVIN
of "Calvin and Hobbes" fame

Leader, bandits at 2 o'clock!

Roger; it's only 1:30 now—what'll I do 'til then?

—THE BILL WATERSON COMIC CHARACTER CALVIN
of "Calvin and Hobbes" fame

It only takes five years to go from rumor to standard operating procedure.

—DICK MARKGRAF

Flying an aeroplane with only a single propeller to keep you in the air. Can you imagine that?

—CAPTAIN JEAN-LUC PICARD
from *Star Trek: The Next Generation* episode "Booby Trap"

MaCleod, since you've flown the SeaBee a lot you'll understand when I say it was the only airplane I ever owned that you could put in a dive, loose a cylinder and stall out!

—Ernest K. Gann

This is an especially good time for you vacationers who plan to fly, because the Reagan administration, as part of the same policy under which it recently sold Yellowstone National Park to Wayne Newton, has "deregulated" the airline industry. What this means for you, the consumer, is that the airlines are no longer required to follow any rules whatsoever. They can show snuff movies. They can charge for oxygen. They can hire pilots right out of Vending Machine Refill Person School. They can conserve fuel by ejecting husky passengers over water. They can ram competing planes in mid-air. These innovations have resulted in tremendous cost savings which have been passed along to you, the consumer, in the form of flights with amazingly low fares, such as $29. Of course, certain restrictions do apply, the main one being that all these flights take you to Newark, and you must pay thousands of dollars if you want to fly back out.

—Dave Barry,
*Iowa—Land of Secure Vacations*

As you know, birds do not have sexual organs because they would interfere with flight. [In fact, this was the big breakthrough for the Wright Brothers. They were watching birds one day, trying to figure out how to get their crude machine to fly, when suddenly it dawned on Wilbur. "Orville," he said, "all we have to do is remove the sexual organs!" You should have seen their original design.] As a result, birds are very, very difficult to arouse sexually. You almost never see an aroused bird. So when they want to reproduce, birds fly up and stand on telephone lines, where they monitor telephone conversations with their feet. When they find a conversation in which people are talking dirty, they grip the line very tightly until they are both highly aroused, at which point the female gets pregnant.

—Dave Barry
*Sex and the Single Amoebae*

Our headline ran, "Virgin screws British Airways." We'd have rather preferred 'British Airways screws Virgin,' but we had to run with the facts.

—NEWS EDITOR
*The Sun* newspaper

Firewall: (1) The part of the airplane specially designed to allow all heat and exhaust to enter the cockpit. (2) The act of pulling 69 inches of manifold pressure, out of an engine designed to pull 60.

—BOB STEVENS
*There I Was*
1968

Now I know what a dog feels like watching TV.

—A DC-9 CAPTAIN TRAINEE
*attempting to check out on the "glass" A-320*

And this, ladies and gentlemen, is the very first Fokker airplane built in the world. The Dutch call it the mother Fokker.

—CUSTODIAN AT THE AVIODOME AVIATION MUSEUM, SCHIPHOL AIRPORT AMSTERDAM

Flight Reservation Systems decide whether or not you exist. If your information isn't in their database, then you simply don't get to go anywhere.

—ARTHUR MILLER

There are only two reasons to sit in the back row of an airplane: Either you have diarrhea, or you're anxious to meet people who do.

—HENRY KISSINGER

Most of us [test pilots] agreed the Cutlass [Chance-Vought F7U-3] could be made into a pretty good flying machine with a few modifications, like adding a conventional tail, tripling the thrust, cutting the nosewheel strut in half, completely redoing the flight control system, and getting someone else to fly it.

—JOHN MOORE
*The Wrong Stuff: Flying on the Edge of Disaster*
1997

I never liked riding in helicopters because there's a fair probability that the bottom part will get going around as fast as the top part.

—LT. COL. JOHN WITTENBORN, USAFR

It looked like a Taco Bell after an earthquake.

—KAREN BRESLAU
reporter for *Newsweek*
*describing Air Force One after hitting severe air turbulence while serving Mexican food*
1996

If black boxes survive air crashes—why don't they make the whole plane out of that stuff?

—GEORGE CARLIN

You land a million planes safely, then you have one little mid-air and you never hear the end of it.

—AIR TRAFFIC CONTROLLER, NEW YORK TRACON
Opening quotation in movie *Pushing Tin*
1999

If God had intended man to fly, He would not have invented Spanish Air Traffic Control.

—DAVE LISTER
in the BBC TV series, *Red Dwarf*

Real planes use only a single stick to fly. This is why bulldozers & helicopters—in that order—need two.

—PAUL SLATTERY

The entrance to the cockpit of this aircraft is most difficult. It should have been made impossible.

—FLIGHT JOURNAL MAGAZINE
*regards the XF10F-1, Grumman's first attempt at a swing-wing fighter*
April 2000

They're beeping and they're flashing. They're flashing and they're beeping! I can't stand it anymore, they're blinking and they're flashing.

—Buck Murdock
**in the 1982 movie *Airplane II, The Sequel***

[When asked why he was referred to as "Ace"]:
Because during World War Two I was responsible for the destruction of six aircraft, fortunately three were enemy.

—Captain Ray Lancaster, USAAF

People think it would be fun to be a bird because you could fly. But they forget the negative side, which is the preening.

—Jack Handey
**Deep Thoughts from *Saturday Night Live***

The light at the end of the tunnel is another airplane's landing light coming down head-on to the runway you are taking off from.

—Robert Livingston
***Flying the Aeronca***
**1981**

What is that mountain goat doing way up here in the clouds?

—Gary Larson
**in a well-known "Farside" cartoon**

Somebody said that carrier pilots were the best in the world, and they must be or there wouldn't be any of them left alive.

—Ernie Pyle

Young man at EAA Oshkosh: "What color are your parachutes?"

Ron Terry, aerobatic pilot: "I don't know and I hope I never find out!"

When I grow up I want to be a pilot because it's a fun job and easy to do.

That's why there are so many pilots flying around these days.

Pilots don't need much school.

They just have to learn to read numbers so they can read their instruments.

I guess they should be able to read a road map, too.

Pilots should be brave so they won't get scared if it's foggy and they can't see, or if a wing or motor falls off.

Pilots have to have good eyes to see through the clouds, and they can't be afraid of thunder or lightning because they are much closer to them than we are.

The salary pilots make is another thing I like.

They make more money than they know what to do with.

This is because most people think that flying a plane is dangerous, except pilots don't because they know how easy it is.

I hope I don't get airsick because I get carsick and if I get airsick, I couldn't be a pilot and then I would have to go to work.

—PURPORTED TO HAVE BEEN WRITTEN BY A FIFTH GRADE STUDENT AT JEFFERSON SCHOOL, BEAUFORT, SC
**It was first published in the *South Carolina Aviation News***

# Clichés

CAUTION: Aviation may be hazardous to your wealth.

If God had wanted me to fly, he would have made me flush riveted.

The average pilot, despite the sometimes swaggering exterior, is very much capable of such feelings as love, affection, intimacy and caring. These feelings just don't involve anyone else.

Without fuel, pilots become pedestrians.

Flying the airplane is more important than radioing your plight to a person on the ground incapable of understanding it.

It is far better to arrive late in this world than early in the next.

Flying helicopters is like masturbating. It feels good while you're doing it, but you're ashamed to tell anyone afterwards.

Eagles may soar, but weasels never get sucked into jet air intakes

If God had meant for men to fly he would have made their bones hollow and not their heads.

*Every redneck cropduster's last words:*

Hey, everybody—watch this!

*The three biggest lies in Army aviation:*

1. You're the only crewmember available.
2. Don't ask me; I'm not the regular crewchief.
3. Wait right here, Sir. The crew bus is on its way.

*The similarity between air traffic controllers and pilots?*

If a pilot screws up, the pilot dies.
If ATC screws up, the pilot dies.

*The difference between a duck and a co-pilot?*

The duck can fly.

Helicopters are really a bunch of parts flying in relatively close formation; all rotating around a different axis. Things work well until one of the parts breaks formation.

Basic Flying Instructions:
1. Try to stay in the middle of the air.
2. Do not go near the edges of it.
3. The edges of the air can be recognized by the appearance of ground, buildings, sea, trees and interstellar space. It is much more difficult to fly there.

Same way, same day.
—NAVY PILOTS
*regards Air Force formation flying skills*

Fighter pilots make movies, attack pilots make history.

*AVIATION DICTIONARY*

Airspeed: Speed of an airplane. Deduct 25% when listening to a Navy pilot.

Bank: The folks who hold the lien on most pilots' cars.

Cone of Confusion: An area about the size of New Jersey, located near the final approach beacon at an airport.

Crab: The squadron Ops Officer.

Dead Reckoning: You reckon correctly, or you are.

Engine Failure: A condition which occurs when all fuel tanks mysteriously become filled with air.

Firewall: Section of the aircraft specially designed to let heat and smoke enter the cockpit.

Glide Distance: Half the distance from the airplane to the nearest emergency landing field.

Hydroplane: An airplane designed to land on a 20,000-foot-long wet runway.

IFR: A method of flying by needle and ripcord.

Lean Mixture: Nonalcoholic beer

Nanosecond: Time delay built into the stall warning system.

Parasitic Drag: A pilot who bums a ride and complains about the service.

Range: Usually about 30 miles beyond the point where all fuel tanks fill with air.

Rich Mixture: What you order at the other guy's promotion party.

Roger: Used when you're not sure what else to say.

Service Ceiling: Altitude at which cabin crews can serve drinks.

Spoilers: The Federal Aviation Administration.

Stall - Technique used to explain to the bank why you car payment is late.

*P = THE PROBLEM LOGGED BY THE PILOT.*
*S = THE SOLUTION LOGGED BY THE MECHANIC.*

P: Left inside main tire almost needs replacement.
S: Almost replaced left inside main tire.

P: Test flight OK, except auto-land very rough.
S: Auto-land not installed on this aircraft.

P: No. 2 propeller seeping prop fluid.
S: No. 2 propeller seepage normal. Nos. 1, 3 and 4 propellers lack normal seepage.

P: Something loose in cockpit.
S: Something tightened in cockpit.

P: Dead bugs on windshield.
S: Live bugs on backorder.

P: Autopilot in "altitude-hold" mode produces a 200-fpm descent.
S: Cannot reproduce problem on ground.

P: Evidence of leak on right main landing gear.

S: Evidence removed.

P: DME volume unbelievably loud.

S: DME volume set to more believable level.

P: Friction locks cause throttle levers to stick.

S: That's what they're there for!

P: Transponder inoperative.

S: Transponder always inoperative in OFF mode.

P: Suspected crack in windscreen.

S: Suspect you're right.

P: Number 3 engine missing.

S: Engine found on right wing after brief search.

P: Aircraft handles funny.

S: Aircraft warned to straighten up, fly right, and be serious.

P: Radar hums.

S: Reprogrammed radar with words.

P: Mouse in cockpit.

S: Cat installed.

# Miscellaneous

FDC 1/9731 FDC SPECIAL NOTICE - DUE TO EXTRADORDINARY CIRCUMSTANCES AND FOR REASONS OF SAFETY. ATTENTION ALL AIRCRAFT OPERATORS, BY ORDER OF THE FEDERAL AVATION COMMAND CENTER, ALL AIRPORTS/AIRDROMES ARE NOT AUTHORIZED FOR LANDING AND TAKEOFF. ALL TRAFFIC INCLUDING AIRBORNE AIRCRAFT ARE ENCOURAGED TO LAND SHORTLY.

—*First official emergency notice issued by the FAA on the morning of September 11, 2001 which grounded all U.S. flight operations until further notice after the simultaneous quadruple hijacking. Aviation was spelled incorrectly, understandable under the conditions*

I believe that my course in sending our Kitty Hawk machine to a foreign museum is the only way of correcting the history of the flying machine, which by false and misleading statements has been perverted by the Smithsonian Institution. In its campaign to discredit others in the flying art, the Smithsonian has issued scores of these false and misleading statements…In a foreign museum this machine will be a constant reminder of the reasons for its being there, and after the people and petty jealousies of the day are gone, the historians of the future may examine the evidence impartially and make history accord with it. Your regret that this old machine must leave the country can hardly be so great as my own.

—Orville Wright
**letter to the Smithsonian**
*regarding sending The Flyer to the Science Museum London, England*

The original Wright brothers aeroplane the world's first power-driven, heavier-than-air machine in which man made free, controlled, and sustained flight invented and built by Wilbur and Orville Wright flown by them at Kitty Hawk, North Carolina December 17, 1903. By original scientific research the Wright brothers discovered the principles of human flight as inventors, builders, and flyers they further developed the aeroplane, taught man to fly, and opened the era of aviation.

—Inscription next to The Flyer
*when it was finally brought back to the United States and unveiled at the Smithsonian in 1948, after that institution dropped claims that Langley was first with powered flight*

In the early days it was fun to fly. You could soar over rooftops and trees, or drop down to meet a passing train and wave at the engineer. The whole sky belonged to you. Now there are so many regulations. The sky is crowded. All the fun is gone.

—KATHERINE STINSON
quoted in *Katherine Stinson: The Flying Schoolgirl*
Debra L. Winegarten
2000

Of course, there is a certain element of danger in flying, as there is in every sport. It is still a question in the minds of those who have tried both flying and motoring if the aerodrome, at its average gait of 38 miles an hour, is not a safer vehicle than an automobile when it goes tearing up the road at the same rate of speed.

—EDITORIAL IN *OUTING*
May 1909

I must place on record my regret that the human race ever learned to fly.

—SIR WINSTON CHURCHILL

He bores me. He ought to have stuck to his flying machine.

—PIERRE AUGUSTE RENOIR
*regards Leonardo Da Vinci*

You little fool! Don't you know it is even dangerous to look at an airplane?

—SPENCER TRACY
advice to Myrna Loy in the 1938 movie *Test Pilot*

It [flying] is not a bad sport, but there's no place to go.

—GLENN H CURTISS
1907

The Admiralty said it was a plane and not a boat, the Royal Air Force said it was a boat and not a plane, the Army were plain not interested.

—SIR CHRISTOPHER COCKERELL
*regards his invention the hovercraft*

No one can realize how substantial the air is, until he feels its supporting power beneath him. It inspires confidence at once.

—OTTO LILIENTHAL

What? Only sixteen hours! Are you sure?

—ORVILLE WRIGHT
*on hearing about the first nonstop flight across the Atlantic*
**15 June 1919**

The Wright brothers' design ... allowed them to survive long enough to learn how to fly.

—MICHAEL POTTS, BEECH AIRCRAFT SPOKESMAN
**regards the Wright wing, the *New York Times***
**17 April 1984**

Can build plane . . . Delivery about three months.

—DONALD HALL
**Chief engineer, Ryan Airlines**
*to Charles Lindbergh's request for feasibility of the airplane later known as* "The Spirit of St. Louis"

Simplicate and Add Lightness

—DESIGN PHILOSOPHY OF ED HEINEMANN
**Douglas Aircraft**

Some fear flutter because they do not understand it. And some fear it because they do.

—THEODORE VON KARMAN, AERODYNAMICIST

I think we can build a better plane.

—WILLIAM BOEING
**The Boeing Company**
**later the company's motto**
**1914**

Engineering is the science of doing things over again.

—JOHN E. (JACK) STEINER
**Boeing 727 Chief Engineer**

We have focused on derivatives for several years, but when it's time to do a new airplane, it's time to do a new airplane.

—MICHAEL B. BAIR
**Boeing Commercial Airplanes vice president for business strategy and development**
*announcing the "sonic cruiser"*
**29 March, 2001**

We badly need an aircraft which will provide the DC-3's reliability, its same ease of maintenance, and a similar low cost. One approach could be to marry a modern turboprop engine to a modern airframe. Surely our design capabilities are great enough to create a plane as advanced . . . as the DC-3 was in its day

—U.S. SENATOR A.S. 'MIKE' MONRONEY
*This former chairman of the Senate Aviation Subcommittee isn't the only one to have this thought, lots of planes have claimed to be "the next DC-3." None have succeeded*

Hey Ridley, that Machometer is acting screwy. It just went off the scale on me.

—GENERAL CHARLES "CHUCK" YEAGER
*first radio transmission after going supersonic for the first time, a coded message indicating success*
**1947**

At 42,000' in approximately level flight, a third cylinder was turned on. Acceleration was rapid and speed increased to .98 Mach. The needle of the machmeter fluctuated at this reading momentarily, then passed off the scale. Assuming that the off-scale reading remained linear, it is estimated that 1.05 Mach was attained at this time.

—(THEN) CAPTAIN CHARLES "CHUCK" YEAGER, AIR CORPS
*formal typewritten test flight report on first supersonic flight*
*NACA tracking data and the XS-1's own oscillograph instrumentation later showed "Glamorous Glennis" has attained Mach 1.06 at about 43,000 feet*
**14 October 1947**

It doesn't look nearly as big as it did the first time I saw one. Mickey McGuire and I used to sit hour after hour in the cockpit of the one that American used for training, at the company school in Chicago, saying to each other, "My God, do you think we'll ever learn to fly anything this big?"

—Ernest K. Gann
*regards the DC-3, quoted in* Flying *magazine*
September 1977

I came to admire this machine which could lift virtually any load strapped to its back and carry it anywhere in any weather, safely and dependably. The C-47 groaned, it protested, it rattled, it leaked oil, it ran hot, it ran cold, it ran rough, it staggered along on hot days and scared you half to death, its wings flexed and twisted in a horrifying manner, it sank back to earth with a great sigh of relief - but it flew and it flew and it flew.

—Len Morgan
*The C-47 was the U.S. military designation for the DC-3*

I didn't start out to chart the skies; it's just no one had done it before.

—Elrey B. Jeppesen
*Captain Jeppesen drew the first approach charts to airports, and founded the company that now supplies them to airlines around the world*

I didn't start the business to make a pile of money. I did it to preserve myself for old age.

—Elrey B. Jeppesen

To the IFR cognoscente, it's a serious misunderstanding of instrument flying to think of an approach plate as a mere map for dropping out of the clouds in search of a runway, at the very least, a plate is a work of art and for the true zealot, it's a symbol of man's continuing struggle against the forces of nature.

—Paul Bertorelli
*IFR* magazine editor

I am drawn to the new chart with all of its colorful intricacies as a gourmet must anticipate the details of a feast . . . I shall keep them forever. As stunning exciting proof that a proper mixture of science and art is not only possible but a blessed union.

—Ernest K. Gann
*Fate Is the Hunter*

Oh, pilot! 'tis a fearful night,
There's danger on the deep.

—Thomas Haynes Bayly
*The Pilot*

It is about a period in aviation which is now gone, but which was probably more interesting than any the future will bring. As time passes, the perfection of machinery tends to insulate man from contact with the elements in which he lives. The "stratosphere" planes of the future will cross the ocean without any sense of the water below. Like a train tunneling through a mountain, they will be aloof from both the problems and the beauty of the earth's surface. Only the vibration from the engines will impress the senses of the traveller with his movement through the air. Wind and heat and moonlight take-offs will be of no concern to the transatlantic passenger. His only contact with these elements will lie in accounts such as this.

—Charles A. Lindbergh
foreword to *Listen! The Wind*
1938

It may be merely the impatience of a woman, but is it not time we ceased to quibble over the exact amount in pounds, shillings, and pence each unit is to contribute to the cost of an All-red route and looked at the broader Imperial aspect? Trade, they used to say, followed the Flag. Today and in the future it will also follow the aerodrome, for without speedy communications commercial competition is impossible.

—Lady Sophie Mary Heath

Newspapers found grand material for front-page stories. The lone fight of human endurance against Nature's overwhelming odds was the favourite. Setting off unknown to face the unknown, against parental opposition, with no money, friends, or influence, ran it a close second. Clichés like "blazing trails," flying over "shark-infected seas," "battling with monsoons," and "forced landings amongst savage tribes" became familiar diet for breakfast.

Unknown names became household words, whilst others, those of the failures, were forgotten utterly except by kith and kin.

—Amy Johnson

That's the best way to cross the Atlantic.

—Sir Arthur Whitten Brown
*first nonstop across the Atlantic, upon landing*
**15 June 1919**

He did it alone. We had a cast of a million.

—Neil Armstrong
*regards Charles A. Lindbergh*

Man's flight through life is sustained by the power of his knowledge.

—Austin "Dusty" Miller
*the quote on the Eagle & Fledgling statue at the U.S. Air Force Academy. The statue was donated by personnel from Air Training Command in 1952*

The time has arrived for my family to give back to America part of the reward that aviation has been instrumental in creating. . . . I hope the Dulles Center will introduce children to the same love for aviation that I have. . . . An airplane rising into the sky is the only hope, the only way to reach into a bigger world.

—Steven Udvar-Hazy
**Founder of International Lease Finance Corporation**
*regards his $60 million donation to help build the Dulles expansion of the National Air and Space Museum. When he was 12 his family fled to the United States following the Soviet invasion of Hungary*
**7 October 1999**

The Wright brothers flew through the smoke screen of impossibility.

—DOROTHEA BRANDE

What of the Wright boys in Dayton? Just around the corner they had a shop and did a bicycle business—and they wanted to fly for the sake of flying. They were Man the Seeker, Man on a Quest. Money was their last thought, their final absent-minded idea. They threw out a lot of old mistaken measurements and figured new ones that stood up when they took off and held the air and steered a course. They proved that "the faster you go the less power you need." One of them died and was laid away under blossoms dropped from zooming planes. The other lived on to meditate: what is attraction? when will we learn why things go when they go? what and where is the power?

—CARL SANDBURG
*The People, Yes*
1932

There was a demon that lived in the air. They said whoever challenged him would die. His controls would freeze up, his plane would buffet wildly, and he would disintegrate. The demon lived at Mach 1 on the meter, seven hundred and fifty miles an hour, where the air could no longer move out of the way. He lived behind a barrier through which they said no man would ever pass. They called it the sound barrier.

—COL. JACK RIDLEY
according to the 1983 movie *The Right Stuff*

I have been luckier than the law of averages should allow. I could never be so lucky again.

—GENERAL JAMES H. DOOLITTLE
from his autobiography, *I Could Never Be So Lucky Again*
1991

Pilot Jack Waddell eased throttles forward; Co-Pilot Brian Wygle called out speeds as a gentle giant of the air began to move; Flight Engineer Jess Wallick kept eyes glued to the gauges. The Boeing Model 747 Superjet gathered speed. The nose lifted. After 4,300 ft—less than half the 9,000 ft runway—main gear of the plane left the concrete. At 11:34 a.m., with a speed of 164 miles an hour, quietly and almost serenely, the age of spacious jets began.

—BOEING MAGAZINE
*first flight of the B-747*

Kansas City Center, this is Air Force One. Would you please change our call sign to SAM 27000.

—COLONEL RALPH ALBERTAZZIE
*39,000 feet over Missouri after hearing that passenger Richard Nixon was no longer President*
**9 August 1974**

Center replied, "Roger SAM 27000, give Mr. Nixon our best."

*Dick Rutan:* Edwards Tower, this is Voyager One. We're ready to go.

*Edwards Tower:* Roger, Voyager One. You're cleared for takeoff. ATC clears Voyager One from Edwards Air Force Base to Edwards Air Force Base via flight plan route. Maintain 8,000. Godspeed.

—DICK RUTAN AND ATC
*starting the first around the world nonstop and non-refueled flight*

*Myanmar Air traffic control:* Hotel Bravo-Bravo Romeo Alpha, what is your departure point and destination?

*Brian Jones:* Departure point, Château d'Oex, Switzerland. Destination, somewhere in northern Africa.

*Myanmar Air traffic control, after several seconds' silence:* If you're going from Switzerland to northern Africa, what in hell are you doing in Myanmar?

—BRIAN JONES AND UNKNOWN CONTROLLER
*approaching Myanmar's air space during record around the world in a balloon flight*
**9 March 1999**

Oh no, it wasn't the aeroplanes. It was Beauty killed the Beast.

—JAMES CREELMAN
**Final words of the movie, *King Kong* (1933 version)**

But you the pathways of the sky
Found first, and tasted heavenly springs,
Unfettered as the lark that sings,
And knew strange raptures—though we sigh,
"Poor Iccarus!"

—FLORENCE EARLE COATES
*Poor Icarus*

Look boys I ain't much of a hand at making speeches, but I got a pretty fair idea that something doggone important is goin' on back there. And I got a fair idea the kinda personal emotions that some of you fellas may be thinkin'. Heck, I reckon you wouldn't even be human beings if you didn't have some pretty strong personal feelings about nuclear combat. I want you to remember one thing, the folks back at home are counting on you and by golly we ain't about to let them down. I tell you something else, if this thing turns out to be half as important as I figure it just might be, I'd say that you're all in line for some important promotions and personal citations when this thing is over with. That goes for ever' last one of you regardless of your race, color or creed. Now let's get this thing on the hump. . .we got some flying to do.

—MAJOR T. J. "KING" KONG
in the 1963 movie *Dr. Strangelove or: How I Learned to Stop Worrying and Love the Bomb*

And if you screw up just this much, you'll be flying a cargo plane full of rubber dog shit out of Hong Kong.

—AIR BOSS JOHNSON
in the 1986 movie *Top Gun*

*Maverick*: I feel the need. . .

*Maverick & Goose together*: The need for speed.

—IN THE 1986 MOVIE *TOP GUN*

*Charlie:* Excuse me Lieutenant. Is there something wrong?

*Maverick:* Yes ma'am. The data on the MIG is inaccurate.

*Charlie:* How's that Lieutenant?

*Maverick:* Well I just happened to see a MIG-28 do. . .

*Goose:* We. . . we.

*Maverick:* Sorry Goose We happened to see a MIG-28 do a 4G negative dive.

*Charlie:* Where did you see this?

*Maverick:* That's classified.

*Charlie:* That's what?

*Maverick:* That's classified. I could tell you, but then I'd have to kill you.

—THE 1986 MOVIE *TOP GUN*

If forced to travel on an airplane, try and get in the cabin with the Captain, so you can keep an eye on him and nudge him if he falls asleep or point out any mountains looming up ahead.

—MIKE HARDING
*The Armchair Anarchist's Almanac*
1981

The bluebird carries the sky on his back.

—HENRY DAVID THOREAU
Journal

In blossom today, then scattered:
Life is so like a delicate flower.
How can one expect the fragrance
To last forever?

—VICE ADMIRAL OHNISHI
Kamikaze Special Attack Force

I put the sweat of my life into this project, and if it's a failure, I'll leave the country and never come back.

—HOWARD HUGHES
*to a U.S. Senate subcommittee regards the HK-1 Hughes Flying Boat aka the "Spruce Goose"*
1946

Though, as he was torn into a pink upper air, she was a good craft to ride in, for her belly was firm and her breasts enabled a flying man good hold and emotions of heady safety. . . . Steering her peasant tits he bounded off stars.

—THOMAS KENEALLY
*Blood Red, Sister Rose: A Novel of the Maid of Orleans*
1974

Nobody kicks ass without tanker gas. Nobody.

—ATTRIBUTED TO THE SPECIAL OPERATIONS KC-135 TANKER CREWS OF PLATTSBURGH AND GRISSOM AFBS IN THE MID 1980S

It's time for the human race to enter the solar system.

—VICE PRESIDENT OF THE U.S. DAN QUAYLE

The Wright Brothers created the single greatest cultural force since the invention of writing. The airplane became the first World Wide Web, bringing people, languages, ideas, and values together.

—BILL GATES, CEO, MICROSOFT CORPORATION

If I had a buck for every time I heard "I always wanted to be a pilot," but lost a buck when hearing the expression, "I wish I spent more time with my kids," how rich am I?

—JEFF "THE CAT" MORRIS
Airline Captain

I've got the greatest job in the world. Northwest sends me to New York ten times a month to have dinner. I've just got to take 187 people with me whenever I go.

—COLIN SOUCY
Northwest Airlines pilot

Chicks dig us, and guys think we're cool.

—TOM KRIZEK
Airline Captain

Air Canada. That's a good name for a Canadian airline.

—JOHNNY CARSON
December 1974

When you think about flying, it's nuts really. Here you are at about 40,000 feet, screaming along at 700 miles an hour and you're sitting there drinking Diet Pepsi and eating peanuts. It just doesn't make any sense.

—David Letterman

Americans have an abiding belief in their ability to control reality by purely material means. . . . airline insurance replaces the fear of death with the comforting prospect of cash.

—Cecil Beaton
*It Gives Me Great Pleasure*
1955

We'd have more luck playing pick-up sticks with our butt-cheeks than we will getting a flight out of here before daybreak.

—Del Griffith (John Candy)
in the 1987 movie *Planes, Trains & Automobiles*

The whole history of the Canadian North can be divided into two periods—before and after the aeroplane.

—Hugh L. Keenleyside
Deputy Canadian Minister of Mines and Resources
October 1949

Air show? Buzz-cut Alabamians spewing colored smoke from their whiz jets to the strains of "Rock You Like A Hurricane"? What kind of countrified rube is still impressed by that?

—Sideshow Bob
on TV's *The Simpsons*
episode "Sideshow Bob's Last Gleaming"
written by Spike Ferensten

The simple expression "Suck, Squeeze, Bang and Blow" is the best way to remember the working cycle of the gas turbine.

—Rolls Royce training manual
2002

Twenty years for 270 murders is less than a month per victim. It's just not right.

—Peter Lowenstein
*father of a young American killed in the 1988 bombing of Pan Am Flight 103, on the conviction of Libyan intelligence official Ali Mohmed al-Megrahi*
February 2001

I know planes, but I don't know INS.

—Berton Beach
**Vice President of Operations Aeroservice Aviation Center, Miami**
*(where suspected 9-11 hijacker Ziad Jarrah studied) regards how Saddam Hussein's stepson Mohammad Saffi was able to enroll without the required student visa.*
**July 2002**

Charles A. Lindbergh
Born 1902, Michigan
Died 1974, Maui

" . . . If I take the wings of the morning,
and dwell in the innermost parts of the sea."
C. A. L.

—The unadorned, flat to the ground gravestone of Charles A. Lindbergh
*He died of cancer on the island of Maui, Hawaii, on 26 August, 1974. He was buried three hours later in simple work clothes*

Off we go into the wild blue yonder,
Climbing high into the sun;
Here they come zooming to meet our thunder,
At 'em boys, Give 'er the gun! (Give 'er the gun now!)
Down we dive, spouting our flame from under,
Off with one helluva roar!
We live in fame or go down in flame. Hey!
Nothing'll stop the U.S. Air Force!

Additional verses:

Minds of men fashioned a crate of thunder,
Sent it high into the blue;
Hands of men blasted the world asunder;
How they lived God only knew! (God only knew then!)
Souls of men dreaming of skies to conquer
Gave us wings, ever to soar!
With scouts before and bombers galore. Hey!
Nothing'll stop the U.S. Air Force!

Bridge: "A Toast to the Host"

Here's a toast to the host
Of those who love the vastness of the sky,
To a friend we send a message of his brother men who fly.
We drink to those who gave their all of old,
Then down we roar to score the rainbow's pot of gold.
A toast to the host of men we boast, the U.S. Air Force!
Zoom!
Off we go into the wild sky yonder,
Keep the wings level and true;
If you'd live to be a grey-haired wonder
Keep the nose out of the blue! (Out of the blue, boy!)
Flying men, guarding the nation's border,
We'll be there, followed by more!
In echelon we carry on. Hey!
Nothing'll stop the U.S. Air Force!

—Robert Crawford

*He didn't write "Hey!"; he actually wrote "SHOUT!" without specifying the word to be shouted. Wherever they appear, the words "U.S. Air Force" have been changed from the original "Army Air Corps." Words in parentheses are spoken, not sung.*

*In 1938, Liberty magazine sponsored a contest for a spirited, enduring musical composition to become the official Army Air Corps song. Of 757 scores submitted, Robert Crawford's was selected by a committee of Air Force wives. The song was officially introduced at the Cleveland Air Races on 2 September 1939.*

*Fittingly, Crawford sang in its first public performance.*

*The first page of the score, which Crawford submitted to the selection committee in July 1939, was carried to the surface of the moon on 30 July 1971 aboard the Apollo 15 'Falcon' lunar module by Colonel David R. Scott and Lieutenant Colonel James B. Irwin. Interestingly, at the moment the 'Falcon' blasted off the surface of the moon with Scott and Irwin on board, a rendition of the 'Air Force Song' was broadcast to the world by Major Alfred M. Worden, who had a tape player aboard the 'Endeavor' command module which was in orbit around the moon. Scott, Irwin and Worden comprised the only all-air-force Apollo crew.*

Fate has ordained that the men who went to the moon to explore in peace will stay on the moon to rest in peace. These brave men, Neil Armstrong and Edwin Aldrin, know that there is no hope for their recovery. But they also know that there is hope for mankind in their sacrifice.

These two men are laying down their lives in mankind's most noble goal: the search for truth and understanding. They will be mourned by their families and friends; they will be mourned by their nation; they will be mourned by the people of the world; they will be mourned by a Mother Earth that dared send two of her sons into the unknown.

In their exploration, they stirred the people of the world to feel as one; in their sacrifice, they bind more tightly the brotherhood of man. In ancient days, men looked at stars and saw their heroes in the constellations. In modern times, we do much the same, but our heroes are epic men of flesh and blood.

Others will follow, and surely find their way home. Man's search will not be denied. But these men were the first, and they will remain the foremost in our hearts. For every human being who looks up at the moon in the nights to come will know that there is some corner of another world that is forever mankind.

—William Safire

*Part of the speech drafted for President Richard M. Nixon to be given to the nation should Neil and Buzz not be able to rejoin the command module and be faced with death on, or around, the moon*

Thousands of volumes have been written about aviation, but we do not automatically have thousands of true and special friends in their authors. That rare writer who comes alive on a page does it by giving of himself, by writing of meanings, and not just of fact or of things that have happened to him. The writers of flight who have done this are usually found together in a special section on private bookshelves.

—Richard Bach

"The Pleasure of Their Company," in *Flying* magazine
April 1968

My wings are a thousand books.

—Gill Robb Wilson

# Space Flight

Those who study the stars have God for a teacher.

—TYCHO BRAHE

Witness this new-made world, another Heav'n
From Heaven Gate not farr, founded in view
On the clear Hyaline, the Glassie Sea;
Of amplitude almost immense, with Starr's
Numerous, and every Starr perhaps a world
Of destined habitation.

—JOHN MILTON
*Paradise Lost*

It [the rocket] will free man from his remaining chains, the chains of gravity which still tie him to this planet. It will open to him the gates of heaven.

—DR. WERNHER VON BRAUN

I have a strong feeling about interesting people in space exploration. . . . And the only way it's going to happen is to have some kid fantasize about getting his ray gun, jumping into his spaceship, and flying into outer space.

—GEORGE LUCAS
Creator of *Star Wars*

A sense of the unknown has always lured mankind and the greatest of the unknowns of today is outer space. The terrors, the joys and the sense of accomplishment are epitomized in the space program.

—WILLIAM SHATNER
*who played Captain Kirk in the original Star Trek*

I know that some knowledgeable people fear that although we might be willing to spend a couple of billion dollars in 1958, because we still remember the humiliation of Sputnik last October, next year we will be so preoccupied by color television, or new-style cars, or the beginning of another national election, that we will be unwilling to pay another year's installment on our space conquest bill. For that to happen well, I'd just as soon we didn't start.

—HUGH L. DRYDEN

We all feel that there's a lot more to this thing than just being Number One, though we all want that. The Number One man will be the tool of our close-knit team. We're just getting started here with space programs that will continue as long as man can pick himself up and go. And we're all going to get a chance to make some contribution. There will be a lot of firsts: the first man on a ballistic firing, the first man into orbit, the first man to orbit the moon, the first man to land on the moon. The public enthusiasm in this thing so far has surprised me. If we don't keep moving, maybe the Russians are going to win a few of these blue ribbons.

—L. GORDON "GORDO" COOPER
**_Life_ magazine**
**1959**

Poyekhali (translation, 'Let's Go!')

—YURI A. GAGARIN
*shouted as Vostok 1 lifted off*
**12 April 1961**

I am a friend, comrades, a friend!

—YURI A. GAGARIN
*first words on the ground after first spaceflight, to a woman and a girl nearby*
**12 April 1961**

*The woman replied:* Can it be that you have come from outer space?

*Yuri*: As a matter of fact, I have!

Why don't you fix your little problem and light this candle?

—ALAN B. SHEPARD JR.
**Cape Canaveral Air Station, just prior to the United States' first manned space mission**
*to Mission Control during his four hour sit atop the 10-story, 33-ton Redstone rocket while last-minute problems were being fixed*
**5 May 1961**

You're on your way, Jose!

—DONALD "DEKE" SLAYTON
*at Mission Control, to Alan Shepard at liftoff of Freedom 7, first American in space*
**5 May 1961**

Roger, liftoff, and the clock has started.

—Alan B. Shepard Jr.
*replying, 09:54 EST*
**5 May 1961**

Godspeed, John Glenn.

—Scott Carpenter
*spoken as Friendship 7 lifted off, but not over the ground-to-air circuit and so not heard by John Glenn,*
*The expression comes from 'God Spede you,' or God prosper you, which was a 15th century Middle English expression of good wishes to a person starting a journey*
**20 February 1962**

Zero G and I feel fine.

—John Glenn
*in orbit*
**20 February 1962**

I don't know what you could say about a day in which you have seen four beautiful sunsets.

—John Glenn

That was a real fireball.

—John Glenn
*re-entry*
**20 February 1962**

I am a stranger. I come in peace. Take me to your leader and there will be a massive reward for you in eternity.

—Note carried by John Glenn
*on his historic flight, translated into several languages, for use if he splashed down in the remote South Pacific seas*

The moon is the first milestone on the road to the stars.

—Arthur C. Clarke

The flight was extremely normal . . . for the first 36 seconds then after that got very interesting.

—Pete Conrad
**Apollo 12 commander**
*regards the launch during which two electrical discharges almost ended the mission*

It was a thunderingly beautiful experience—voluptuous, sexual, dangerous, and expensive as hell.

—KURT VONNEGUT, JR.
*Playboy* interview
*regards the Apollo launches*
1973

It means nothing to me. I have no opinion about it, and I don't care.

—PABLO PICASSO
quoted in the *New York Times*
*reacting to the first moon landing*
21 July 1969

Prometheus is reaching out for the stars with an empty grin on his face.

—ARTHUR KOESTLER
the *New York Times*
*regarding the first moon landing*
21 July 1969

Our passionate preoccupation with the sky, the stars, and a God somewhere in outer space is a homing impulse. We are drawn back to where we came from.

—ERIC HOFFER
the *New York Times*
*regards the first moon landing*
21 July 1969

Space is the stature of God.

—JOSEPH JOUBERT
French essayist, moralist
*Pensées*, Chapter 12
1842

Treading the soil of the moon, palpating its pebbles, tasting the panic and splendor of the event, feeling in the pit of one's stomach the separation from terra . . . these form the most romantic sensation an explorer has ever known . . . this is the only thing I can say about the matter. The utilitarian results do not interest me.

—VLADIMIR NABOKOV
quoted in the *New York Times*
*referring to the first moon landing*
21 July 1969

So there he is at last. Man on the moon. The poor magnificent bungler! He can't even get to the office without undergoing the agonies of the damned, but give him a little metal, a few chemicals, some wire and twenty or thirty billion dollars and, vroom! there he is, up on a rock a quarter of a million miles up in the sky.

—Russell Baker
the *New York Times*
21 July 1969

We have seen a wonder. There has never been one quite like it. What first steps in human history would one have chosen to witness, if one could travel in time? The Vikings coming ashore wherever they did come ashore—Newfoundland?—in North America? Or the first little boat from Columbus's ship scraping the land under her keel? Yet all of that, or any other bit of geographical discovery, we should be seeing with hindsight. On the spot, it must have seemed much more down-to-earth. People getting out of boats must have looked (and felt) very much like people getting out of boats anywhere at anytime.

No, we have had the best of it. We have seen something unique. It is right that it should have looked like something we have never seen before. In science films, perhaps—but this was real. The figure, moving so laboriously, as though it was learning, minute by minute, to walk, was a man of our own kind. Inside that gear there was a foot, a human foot. Watch. It has come, probing its way down—near to something solid. One expects to hear (there is no air, one could hear nothing) a sound. At last, it has come down. Onto a surface. Onto the surface of the moon.

Well, we have seen a wonder. We ought to count our blessings.

—Lord C. P. Snow
*Look* magazine
1969

The world is being Americanized and technologized to its limits, and that makes it dull for some people. Reaching the Moon restores the frontier and gives us the lands beyond.

—Isaac Asimov
*regards Apollo program*

When old dreams die, new ones come to take their place. God pity a one-dream man.

—ESTHER GODDARD
*reading from her late husband's diary to the AP just prior to the launch of Apollo 11*

Suddenly I saw a meteor go by underneath me. A moment later I found myself think, that can't be a meteor. Meteors burn up in the atmosphere above us; this was below us. Then, of course, the realization hit me.

—JEFFREY HOFFMAN

*CAPCOM Richard Covey:* Challenger Houston, you are go at Throttle Up.

*Cmdr. Dick Scobee:* Roger Houston, Go at Throttle Up

Pilot Mike Smith: uh-oh . . .

—LAST WORDS RECORDED FROM SPACE SHUTTLE CHALLENGER BEFORE EXPLODING 74 SECONDS INTO ITS FLIGHT
**28 January 1986**

Obviously a major malfunction.

—STEPHEN A. NESBITT
**NASA Public Affairs Officer**
*live on air, just moments after the space shuttle Challenger exploded*
**28 January 1986**

We shall never forget them nor the last time we saw them, as they prepared for their mission and waved good-bye and slipped the surly bonds of Earth to touch the face of God.

—PRESIDENT RONALD REAGAN
*addressing NASA employees following the tragic loss of the Challenger 7 crew on STS-51L*

For those who have seen the Earth from space, and for the hundreds and perhaps thousands more who will, the experience most certainly changes your perspective. The things that we share in our world are far more valuable than those which divide us.

—DONALD WILLIAMS

My first view—a panorama of brilliant deep blue ocean, shot with shades of green and gray and white—was of atolls and clouds. Close to the window I could see that this Pacific scene in motion was rimmed by the great curved limb of the Earth. It had a thin halo of blue held close, and beyond, black space. I held my breath, but something was missing—I felt strangely unfulfilled. Here was a tremendous visual spectacle, but viewed in silence. There was no grand musical accompaniment; no triumphant, inspired sonata or symphony. Each one of us must write the music of this sphere for ourselves.

—Charles Walker

Once during the mission I was asked by ground control what I could see. "What do I see?" I replied. "Half a world to the left, half a world to the right, I can see it all. The Earth is so small."

—Vitali Sevastyanov

Looking outward to the blackness of space, sprinkled with the glory of a universe of lights, I saw majesty—but no welcome. Below was a welcoming planet. There, contained in the thin, moving, incredibly fragile shell of the biosphere is everything that is dear to you, all the human drama and comedy. That's where life is; that's were all the good stuff is.

—Loren Acton

The Earth was small, light blue, and so touchingly alone, our home that must be defended like a holy relic. The Earth was absolutely round. I believe I never knew what the word round meant until I saw Earth from space.

—Aleksei Leonov

We went to the moon as technicians; we returned as humanitarians.

—Edgar Mitchell

The sun truly "comes up like thunder," and it sets just as fast. Each sunrise and sunset lasts only a few seconds. But in that time you see at least eight different bands of color come and go, from a brilliant red to the brightest and deepest blue. And you see sixteen sunrises and sixteen sunsets every day you're in space. No sunrise or sunset is ever the same.

—JOSEPH ALLEN

The powered flight took a total of about eight and a half minutes. It seemed to me it had gone by in a flash. We had gone from sitting still on the launch pad at the Kennedy Space Center to traveling at 17,500 miles an hour in that eight and a half minutes. It is still mind-boggling to me. I recall making some statement on the air-to-ground radio for the benefit of my fellow astronauts, who had also been in the program a long time, that it was well worth the wait.

—BOB CRIPPEN
*regards first flight of the Space Shuttle, STS-1*

The vehicle explodes, literally explodes, off the pad. The simulator shakes you a little bit, but the actual liftoff shakes your entire body and soul.

—MIKE MCCULLEY

The only thing I've experienced that could compare to the launch in terms of raw power was the Loma Prieta earthquake.

—LOREN ACTON

In the press grandstand where I watched Discovery rise against the cloudless sky, the media hit the abort button on cynicism. The Earth shook to the sounds of man, three miles away. The candle lit. . . only someone stripped of awe can leave a launch untouched.

—JONATHAN ALTER
*Newsweek* magazine
9 November, 1998

To fly in space is to see the reality of Earth, alone. The experience changed my life and my attitude toward life itself. I am one of the lucky ones.

—ROBERTA BONDAR

Suddenly, from behind the rim of the moon, in long, slow-motion moments of immense majesty, there emerges a sparkling blue and white jewel, a light, delicate sky-blue sphere laced with slowly swirling veils of white, rising gradually like a small pearl in a thick sea of black mystery. It takes more than a moment to fully realize this is Earth . . . home.

—EDGAR MITCHELL

My view of our planet was a glimpse of divinity.

—EDGAR MITCHELL

For the first time in my life I saw the horizon as a curved line. It was accentuated by a thin seam of dark blue light – our atmosphere. Obviously this was not the ocean of air I had been told it was so many times in my life. I was terrified by its fragile appearance.

—ULF MERBOLD

As you pass from sunlight into darkness and back again every hour and a half, you become startlingly aware how artificial are thousands of boundaries we've created to separate and define. And for the first time in your life you feel in your gut the precious unity of the Earth and all the living things it supports.

—RUSSELL SCHWEIKART
*returning from Apollo 9*

The Lunar landing of the astronauts is more than a step in history; it is a step in evolution.

—THE *NEW YORK TIMES* EDITORIAL
20 July 1969

For when I look at the moon I do not see a hostile, empty world. I see the radiant body where man has taken his first steps into a frontier that will never end.

—DAVID R. SCOTT
Commander Apollo 15
*National Geographic*, Volume 144, No 3
September 1973

The world itself looks cleaner and so much more beautiful. Maybe we can make it that way—the way God intended it to be—by giving everybody that new perspective from out in space.

—ROGER B. CHAFFEE

We have taken to the Moon the wealth of this nation, the vision of its political leaders, the intelligence of its scientists, the dedication of its engineers, the careful craftsmanship of its workers, and the enthusiastic support of its people.

We have brought back rocks, and I think it is a fair trade . . .

Man has always gone where he has been able to go. It's that simple. He will continue pushing back his frontier, no matter how far it may carry him from his homeland.

—MICHAEL COLLINS

What was most significant about the lunar voyage was not that man set foot on the moon but that they set eye on the earth.

—NORMAN COUSINS

Earth bound history has ended. Universal history has begun.

—EARL HUBBARD

A Chinese tale tells of some men sent to harm a young girl who, upon seeing her beauty, become her protectors rather than her violators. That's how I felt seeing the Earth for the first time. I could not help but love and cherish her.

—TAYLOR WANG

I cannot join the space program and restart my life as an astronaut, but this opportunity to connect my abilities as an educator with my interests in history and space is a unique opportunity to fulfill my early fantasies.

—CHRISTA MCAULIFFE
**Teacher**
*from her winning essay in NASA's nationwide search for the first teacher to travel in space, released after her death with six others aboard the space shuttle Challenger*

Our nation is indeed fortunate that we can still draw on an immense reservoir of courage, character, and fortitude, that we are still blessed with heroes like those of the space shuttle Challenger. Man will continue his conquest of space. To reach out for new goals and ever-greater achievements, that is the way we shall commemorate our seven Challenger heroes.

—U.S. President Ronald Reagan

We are all . . . children of this universe. Not just Earth, or Mars, or this system, but the whole grand fireworks. And if we are interested in Mars at all, it is only because we wonder over our past and worry terribly about our possible future.

—Ray Bradbury
*Mars and the Mind of Man*
1973

Man is an artifact designed for space travel. He is not designed to remain in his present biologic state any more than a tadpole is designed to remain a tadpole.

—William Burroughs
*Civilian Defense*
1985

I didn't care if I was first, 50th, or 500th in space. I just wanted to go.

—Dennis Tito, first space tourist
quoted in *USA Today*
*during Mir training*
20 June 2000

We are very happy to accompany you to space. We like your mathematical mind. And we more like your romantic soul.

—Yuri Baturin
cosmonaut flight engineer
*regards Dennis Tito*
*Newsweek*
April 2001

You are not a baby. I am not a babysitter. I am commander. You are "first cosmonaut tourist," an "engineer in education." It is very important for mankind.

—Talgat Musabayev
**cosmonaut commander**
*regards Dennis Tito*
*Newsweek*
**April 2001**

But the astronauts who lost their lives on Challenger, as well as the other eight astronauts who were killed in the line of duty and the four Soviet cosmonauts who died in space serve as inspiration for us all. None of them would have wanted to give her or his life in vain. None would have wanted us to stop striving for the stars. If anything, we must continue to preserve their dreams.

—Doug Fulmer
*Ad Adstra*
**July/August 1991**

Bob, this is Gene, and I'm on the surface; and, as I take man's last step from the surface, back home for some time to come—but we believe not too long into the future—I'd like to just [say] what I believe history will record. That America's challenge of today has forged man's destiny of tomorrow. And, as we leave the Moon at Taurus- Littrow, we leave as we came and, God willing, as we shall return, with peace and hope for all mankind. Godspeed the crew of Apollo 17.

—Eugene Cernan
**Apollo 17 Commander**
*Last man to walk on the moon*
**14 December 1972**

The regret on our side is, they used to say years ago, we are reading about you in science class. Now they say, we are reading about you in history class.

—Neil Armstrong
**July 1999**

Of all investments into the future, the conquest of space demands the greatest efforts and the longest-term commitment . . . but it also offers the greatest reward: none less than a universe.

—Daniel Christlein

A few million years ago there were no humans. Who will be here a few million years hence? In all the 4.6-billion-year history of our planet, nothing much ever left it. But now, tiny unmanned exploratory spacecraft from Earth are moving, glistening and elegant, through the solar system. We have made a preliminary reconnaissance of twenty worlds, among them all of the planets visible to the naked eye, all those wandering nocturnal lights that stirred our ancestors toward understanding and ecstasy. If we survive, our time will be famous for two reasons: that at this dangerous moment of technological adolescence we managed to avoid self-destruction; and because this is the epoch in which we began our journey to the stars.

—Carl Sagan
*Cosmos*
1980

My friends they were dancing here in the streets of Huntsville when our first satellite orbited the Earth. They were dancing again when the first Americans landed on the moon. I'd like to ask you, don't hang up your dancing slippers.

—Dr. Wernher von Braun

# Unidentified Flying Objects

I have argued flying saucers with lots of people. I was interested in this: they keep arguing that it is possible. And that's true. It is possible. They do not appreciate that the problem is not to demonstrate whether it's possible or not but whether it's going on or not.

—Richard Feynman

Mission Control, please be informed, there is a Santa Claus.

—Commander James Lovell

*He made this transmission after coming around the far side of the moon on the Apollo 8 mission. Although it was Christmas time, this statement has caused considerable controversy as "Santa Claus" was supposedly a codeword used to indicate a UFO or other unusual sighting.*

**1968**

[The Airship] looked like a great black cigar with a fishlike tail. . . . The body was at least 100 feet long and attached to it was a triangular tail, one apex being attached to the main body. The surface of the airship looked as if it were made of aluminum, which exposure to wind and weather had turned dark. . . . . The airship went at tremendous speed. As it neared Lorin it turned quickly and disappeared in the direction of San Francisco. At half past 8 we saw it again, when it took about the same direction and disappeared.

—Case Gilson

**reported in the *Oakland Tribune***

**1 December 1896**

It was the darndest thing I've ever seen. It was big, it was very bright, it changed colors and it was about the size of the moon. We watched it for ten minutes, but none of us could figure out what it was. One thing's for sure, I'll never make fun of people who say they've seen unidentified objects in the sky.

If I become President, I'll make every piece of information this country has about UFO sightings available to the public and the scientists.

—President Jimmy Carter

I am completely convinced that [UFOs] have an out-of-world basis.

—Dr. Walther Riedel
*Life* magazine
*Research director and chief designer at Germany's rocket center in Peenemunde. He also worked on classified projects for the U.S. after WWII*
7 April 1952

I can assure you that, given they exist, these flying saucers are made by no power on this Earth.

—President Harry S. Truman
*at a press conference*
4 April 1950

For reasons of national security and out of consideration for some people still alive I have omitted certain material. Some of this material cannot be made available for many years, perhaps for many generations.

—Preface to the Memoirs of President Harry S. Truman
1946–52 Vol II: *Years of Trial and Hope*

Of course the flying saucers are real and they are interplanetary. . . . The cumulative evidence for the existence of UFOs is quite overwhelming and I accept the fact of their existence.

—Air Chief Marshal Lord Hugh Dowding, RAF
August 1954

I believe that these extraterrestrial vehicles and their crews are visiting this planet from other planets—which obviously are a little more technically advanced than we are here on earth.

—Astronaut L. Gordon "Gordo" Cooper
letter to the United Nations
1978

We all know that UFOs are real. All we need to ask is where do they come from.

—Astronaut Edgar Mitchell
*after his Apollo 14 moon flight in 1971*

Flying saucers are real. Too many good men have seen them, that don't have hallucinations.

—Captain Edward "Eddie" Rickenbacker

I've been asked [about UFOs] and I've said publicly I thought they were somebody else, some other civilization.

—Astronaut Eugene Cernan
Apollo 17
in *Los Angeles Times*
1973

I've been convinced for a long time that the flying saucers are real and interplanetary. In other words we are being watched by beings from outer space.

—Albert M. Chop
Deputy Public Relations Director at NASA

I think that it is much more likely that the reports of flying saucers are the results of the known irrational characteristics of terrestrial intelligence than of the unknown rational efforts of extraterrestrial intelligence.

—Richard Feynman

Our journeys to the stars will be made on spaceships created by determined, hardworking scientists and engineers applying the principles of science, not aboard flying saucers piloted by little gray aliens from some other dimension.

—Robert A. Baker
"The Aliens Among Us: Hypnotic Regression Revisited"
*The Skeptical Inquirer*, Winter 1987–88 Vol. 12 (2)

Until they come to see us from their planet, I wait patiently. I hear them saying: Don't call us, we'll call you.

—Marlene Dietrich
*Venus*
1962

A circle of fire coming in the sky, noiseless, one rod long with its body and one rod wide. After some days these things became more numerous, shining more than the brightness of the sun.

—FROM A SERIES OF EGYPTIAN HIEROGLYPHS ON A PAPYRUS DATED TO THE REIGN OF THUTMOSE III
(1504-1450 BC)

The phenomenon of UFOs does exist, and it must be treated seriously.

—MIKHAIL GORBACHEV
*Soviet Youth*
4 May 1990

As far as I know, an alien spacecraft did not crash in Roswell, New Mexico, in 1947. . . . If the United States Air Force did recover alien bodies, they didn't tell me about it either, and I want to know.

—PRESIDENT BILL CLINTON
Reply to a letter from a child asking about the Roswell Incident

Sometimes I think we're alone in the universe, and sometimes I think we're not. In either case the idea is quite staggering.

—ARTHUR C. CLARKE

# Skydiving

Out of 10,000 feet of fall, always remember that the last half an inch hurts the most.

—Captain Charles W. Purcell
1932

If riding in an airplane is flying, then riding in a boat is swimming. If you want to experience the element, then get out of the vehicle.

—Anon

When the people look like ants—Pull.
When the ants look like people—Pray.

—Anon

I now know the color of fear—It's brown.

—Anon

Man small
Why fall?
Skies call
That's all.

—Anon

Only skydivers know why the birds sing.

—Anon

It is one thing to be in the proximity of death, to know more or less what she is, and it is quite another thing to seek her.

—Ernest Hemingway

Skydivers do it in the stable spread position.

—Cliché

Skydiving has been my life, and it will probably be my death too. But hopefully not yet, for I have many years of jumps left in me.

—Robin Wilcox
*four days before dying too soon*
1987

# Birds

The very idea of a bird is a symbol and a suggestion to the poet. A bird seems to be at the top of the scale, so vehement and intense his life. . . . The beautiful vagabonds, endowed with every grace, masters of all climes, and knowing no bounds—how many human aspirations are realised in their free, holiday-lives—and how many suggestions to the poet in their flight and song!

—JOHN BURROUGHS
*Birds and Poets*
1887

When thou seest an eagle, thou seest a portion of genius; lift up thy head!

—WILLIAM BLAKE

Look at that mallard as he floats on the lake; see his elevated head glittering with emerald green, his amber eyes glancing in the light! Even at this distance, he has marked you, and suspects that you bear no goodwill towards him, for he sees that you have a gun, and he has many a time been frightened by its report, or that of some other. The wary bird draws his feet under his body, springs upon then, opens his wings, and with loud quacks bids you farewell.

—JOHN JAMES AUDUBON
*Birds of America*
1840

I wish the bald eagle had not been chosen as the representative of our country; he is a bird of bad moral character; like those among men who live by sharping and robbing, he is generally poor, and often very lousy. The turkey is a much more respectable bird, and withal a true original native of America.

—BENJAMIN FRANKLIN
in a letter to Sarah Bache

*King Henry:* But what a point, my lord, your falcon made, And what a pitch she flew above the rest! To see how God in all his creatures works! Yea, man and birds are fain of climbing high.

*Suffolk:* No marvel, an it like your majesty, My lord protectors hawks do tower so well; They know their masters loves to be aloft, And bears his thoughts above his falcon's pitch.

*Gloucester:* My lord, 'tis but a base ignoble mind That mounts no higher than a bird can soar.

—William Shakespeare
*Henry VI, Part 2*, Act 2

I once had a sparrow alight upon my shoulder for a moment, while I was hoeing in a village garden, and I felt that I was more distinguished by that circumstance that I should have been by any epaulet I could have worn.

—Henry David Thoreau

I hope you love birds too. It is economical. It saves going to heaven.

—Emily Dickinson

Can you imagine any better example of divine creative accomplishment than the consummate flying machine that is a bird? The skeleton, very flexible and strong, is also largely pneumatic—especially in the bigger birds. The beak, skull, feet, and all the other bones of a 25-pound pelican have been found to weigh but 23 ounces.

—Guy Murchie, Jr.
*Song of the Sky*
1954

We are all pirates at heart. There is not one of us who hasn't had a little larceny in his soul. And which one of us wouldn't soar if God had thought there was merit in the idea? So, when we see one of those great widespread pirates soaring across the grain of sea winds we thrill, and we long, and, if we are honest, we curse that we must be men every day. Why not one day a bird! There's an idea, now, one day out of seven a pirate in the sky. What puny power a man can attain by comparison. Compare a 747 with a bird and blush!

—ROGER CARAS
*Birds and Flight*
1971

Lying under an acacia tree with the sound of the dawn around me, I realized more clearly the facts that man should never overlook: that the construction of an airplane, for instance, is simple when compared [with] a bird; that airplanes depend on an advanced civilization, and that where civilization is most advanced, few birds exist. I realized that if I had to choose, I would rather have birds than airplanes.

—CHARLES A. LINDBERGH
*interview shortly before his death*
1974

# Ballooning

How posterity will laugh at us, one way or other! If half a dozen break their necks, and balloonism is exploded, we shall be called fools for having imagined it could be brought to use: if it should be turned to account, we shall be ridiculed for having doubted.

—HORACE WALPOLE
letter to Horace Mann
24 June 1785

The best way of travel, however, if you aren't in any hurry at all, if you don't care where you are going, if you don't like to use your legs, if you don't want to be annoyed at all by any choice of directions, is in a balloon. In a balloon, you can decide only when to start, and usually when to stop. The rest is left entirely to nature.

—WILLIAM PENE DU BOIS
*The Twenty-One Balloons*
1947

These vehicles can serve no use until we can guide them. I had rather now find a medicine that can cure an asthma.

—DR. SAMUEL JOHNSON

Build lightness in.

—ALBERTO SANTOS-DUMONT

Suddenly the wind ceased. The air seemed motionless around us. We were off, going at the speed of the air-current in which we now lived and moved. Indeed, for us there was no more wind; and this is the first great fact of spherical ballooning. Infinitely gentle is this unfelt motion forward and upward. The illusion is complete: it seems not to be the balloon that moves, but the earth that sinks down and away. . .

Villages and woods, meadows and chateaux, pass across the moving scene, out of which the whistling of locomotives throws sharp notes. These faint, piercing sounds, together with the yelping and barking of dogs, are the only noises that reach one through the depths of the upper air. The human voice cannot mount up into these boundless solitudes. Human beings look like ants along the white lines that are highways; and the rows of houses look like children's playthings.

—ALBERTO SANTOS-DUMONT
*My Air-Ships*, New York, The Century Company
1904

The balloon seems to stand still in the air while the earth flies past underneath.

—ALBERTO SANTOS-DUMONT

There's something in a flying horse,
There's something in a huge balloon.

—WILLIAM WORDSWORTH
*Peter Bell*, Prologue, Stanza 1

As we were returning to the inn we beheld something floating in the ample field of golden evening sky, above the chalk cliffs and the trees that grow along their summit. It was too high up, too large, and too steady for a kite; and, as it was dark, it could not be a star. . . The village was dotted with people with their heads in air; and the children were in a bustle all along the street and far up the straight road that climbs the hill, where we could still see them running in loose knots. It was a balloon, we learned, which had left St. Quentin at half past five that evening. Mighty composedly the majority of the grown people took it. But we were English, and were soon running up the hill with the best. Being travelers ourselves in a small way, we would fain have seen these other travelers alight.

The spectacle was over by the time we gained the top of the hill. All the gold had withered out of the sky, and the balloon had disappeared. Whither? I ask myself; caught up into the seventh heaven? or come safely to land somewhere in that blue uneven distance, into which the roadway dipped and melted before our eyes? Probably the aeronauts were already warming themselves at a farm chimney, for they say it is cold in these unhomely regions of the air. The night fell swiftly. Roadside trees and disappointed sight-seers, returning through the meadows, stood out in black against a margin of low, red sunset. It was cheerfully to face the other way, and so down the hill we went, with a full moon, the color of a melon, swinging high above the wooded valley, and the white cliffs behind us faintly reddened by the fire of the chalk kilns.

—ROBERT LOUIS STEVENSON
*An Inland Voyage*
*his travelogue of a canoe trip from Antwerp to Paris, written when he was 25*
1878

I have known today a magnificent intoxication. I have learnt how it feels to be a bird. I have flown. Yes I have flown. I am still astonished at it, still deeply moved.

—*Le Figaro*
1908

The winds have welcomed you with softness,
The sun has greeted you with its warm hands,
You have flown so high and so well,
That God has joined you in laughter,
And set you back gently into
The loving arms of Mother Earth.

—Anon, known as "The Balloonists Prayer," believed to have been adapted from an old Irish sailors' prayer

The way the public sees it is this. If we don't leave, we are idiots. If we do leave but don't succeed in our mission, we are incompetent. But if we do succeed, it's because it was easy and anyone could have done it.

—Bertrand Piccard
*first to balloon around the world*
1999

# Piloting

The greater the difficulty the more glory in surmounting it. Skillful pilots gain their reputation from storms and tempests.

—Epictetus

Now, there are two ways of learning to ride a fractious horse: one is to get on him and learn by actual practice how each motion and trick may be best met; the other is to sit on a fence and watch the beast a while and then retire to the house and at leisure figure out the best way of overcoming his jumps and kicks. The latter system is the safer, but the former, on the whole, turns out the larger proportion of good riders. It is very much the same thing in learning to ride a flying machine.

—Wilbur Wright
from an address to the Western Society of Engineers in Chicago
18 September 1901

Real confidence in the air is bred only by mistakes made and recovered from at a safe altitude, in a safe ship, and seated on a good parachute.

—Rodney H. Jackson
"A Lesson in Stunting," *Aeronautics* magazine
February 1930

In soloing—as in other activities—it is far easier to start something than it is to finish it.

—Amelia Earhart

No matter how important a man at sea may consider himself, unless he is fundamentally worthy the sea will some day find him out.

—Felix Riesenberg

Greater prudence is needed rather than greater skill.

—Wilbur Wright
1901

It is possible to fly without motors, but not without knowledge and skill.

—Wilbur Wright

There are two critical points in every aerial flight—its beginning and its end.

—ALEXANDER GRAHAM BELL
1906

Anybody can jump a motorcycle. The trouble begins when you try to land it.

—EVEL KNIEVEL

When anyone asks me how I can best describe my experiences of nearly forty years at sea, I merely say uneventful. I have never been in an accident of any sort worth speaking about . . . . I never saw a wreck and have never been wrecked, nor was I ever in any predicament that threatened to end in disaster of any sort.

—CAPTAIN EDWARD J. SMITH
*R.M.S. Titanic*
*an experienced 62-year-old captain, regards what was scheduled to be his last voyage prior to retirement*
1912

Never fly in the same cockpit with someone braver than you.

—RICHARD HERMAN JR.
*Firebreak*
1991

Just remember, if you crash because of weather, your funeral will be held on a sunny day.

—LAYTON A. BENNETT

They will pressure you into doing things that may be unsafe, use your good judgment, and remember, "I would rather be laughed at, than cried for."

—GEORGE MACDONALD

After reading . . . accounts . . . of minor accidents of flight, it is little wonder that the average man would far rather watch someone else fly and read of the narrow escapes from death when some pilot has had a forced landing or a blowout, than to ride himself. Even in the postwar days of now obsolete equipment, nearly all of the serious accidents were caused by inexperienced pilots who were then allowed to fly or attempt to fly—without license or restrictions about anything they could coax into the air . . .

—CHARLES A. LINDERGH
*We*
1928

In flying I have learned that carelessness and overconfidence are usually far more dangerous than deliberately accepted risks.

—WILBUR WRIGHT
in a letter to his father
September 1900

I've learned that it is what I do not know that I fear, and I strive, outwardly from pride, inwardly from the knowledge that the unknown is what will finally kill me, to know all there is to be known about my airplane. I will never die.

—RICHARD BACH
*Stranger to the Ground*
1963

It is not only fine feathers that make fine birds.

—AESOP
"The Jay and the Peacock," *Fables*

Remember, you fly an airplane with your head, not your hands and feet.

—BEVO HOWARD

Fly with the head and not with the muscles. That is the way to long life for a fighter pilot. The fighter pilot who is all muscle and no head will never live long enough for a pension.

—COLONEL WILLIE BATS, GAF
*237 Victories, W.W. II*

Flying is so many parts skill, so many parts planning, so many parts maintenance, and so many parts luck. The trick is to reduce the luck by increasing the others.

—DAVID L. BAKER

It doesn't do any good to stand on the airplane's brakes when you're already on your back!

—REX THORP

Harmony comes gradually to a pilot and his plane. The wing does not want so much to fly true as to tug at the hands that guide it; the ship would rather hunt the wind than lay her nose to the horizon far ahead. She has a derelict quality in her character; she toys with freedom and hints at liberation, but yields her own desires gently.

—BERYL MARKHAM
*West with the Night*
1942

I think there is something exhilarating in flying amongst clouds, and always get a feeling of wanting to pit my aeroplane against them, charge at them, climb over them to show them you have them beat, circle round them, and generally play with them; but clouds can on occasion hold their own against the aviator, and many a pilot has found himself emerging from a cloud not on a level keel.

Cloud-flying requires practice, even if you have every modern instrument, and unless you keep calm and collected you will get into trouble after you have been inside a really thick one for a few minutes. In the very early days of aviation, 1912 to be correct, I emerged from a cloud upside down, much to my discomfort, as I didn't know how to get right way up again. I found out somehow, or I wouldn't be writing this.

—CHARLES RUMNEY SAMSON
*A Flight from Cairo to Cape Town and Back*
1931

I sometimes still go out hunting for bad weather, flying low in simple airplanes to explore the inner reaches of the clouds. Less experienced pilots occasionally join me, not to learn formal lessons about weather flying, but with a more advanced purpose in mind—to accompany me in the slow accumulation of experience through circumstances that never repeat in a place that defies mastery.

—William Langewiesche
*Inside the Sky*
1998

When a flight is proceeding incredibly well, something was forgotten.

—Robert Livingston
*Flying The Aeronca*
1981

There are old pilots and there are bold pilots, but there are no old, bold pilots.

—attributed to W.W. Windstaff
circa W.W. I.

I just made a balls of it, old boy. That's all there was to it.

—Group Captain Sir Douglas Bader RAF
*about his December 1931 roll performed immediately after takeoff that ended in the crash that led to the loss of both legs. He later flew again and led a wing of Spitfires during the Battle of Britain*

There is no problem so complex that it cannot simply be blamed on the pilot.

—Dr. Earl Weiner
Human Factors Expert

I could be president of Sikorsky for six months before they found me out, but the president would only have my job for six seconds before he'd kill himself.

—Walter R. "Dick" Faull
Test Pilot

Learning should be fun. If you don't have fun in aviation then you don't learn, and when learning stops, you die.

—Pete Campbell
FAA

Hours and hours passed, with nothing to do but keep the compass on its course and the plane on a level keel. This sounds easy enough, but its very simplicity becomes a danger when your head keeps nodding with weariness and utter boredom and your eyes everlastingly try to shut out the confusing rows of figures in front of you, which will insist on getting jumbled together. Tired of trying to sort them out, you relax for a second, then your head drops and you sit up with a jerk, Where are you? What are you doing here? Oh yes, of course, you are somewhere in the middle of the North Atlantic, with hungry waves below you like vultures impatiently waiting for the end.

—AMY JOHNSON

To be alone in the air at night is to be very much alone indeed. . . cut off from everything and everyone . . . nothing is "familiar" any longer . . . . I think that unfamiliarity is the most difficult thing to face; one feels rather like Alice in Wonderland after she has nibbled the toadstool that made her grow smaller—and like Alice, one hopes that the process will stop while there is still something left!

—PAULINE GOWER

I am not a very timid type. It's very important to some people, but not to me. I have a simple philosophy: worry about those things you can fix. If you can't fix it, don't worry about it; accept it and do the best you can.

—GENERAL JAMES H. DOOLITTLE

I was always afraid of dying. Always. It was my fear that made me learn everything I could about my airplane and my emergency equipment, and kept me flying respectful of my machine and always alert in the cockpit.

—GENERAL CHARLES "CHUCK" YEAGER
*Yeager, an Autobiography*
1985

Mistakes are inevitable in aviation, especially when one is still learning new things. The trick is to not make the mistake that will kill you.

—STEPHEN COONTS

There is only one rule—Rule One—TNB—Trust No Bastard—they are all trying to kill you.

—CAPTAIN RICK DAVIES
**Chief Pilot, Royal Flying Doctor Service of Australia (Queensland Section)**
*advice given to new captains*

Better to hit the far fence at ten knots than the close fence at $V_{ref}$.

—CAPTAIN RICK DAVIES
**Chief Pilot, Royal Flying Doctor Service of Australia (Queensland Section)**
*advice given to new captains*

You can't lomcevak in an F-16, but you can't go Mach in a Pitts.

—ED HAMILL
**who has flown both aircraft**

"Are you ever afraid when you fly?"

"That's a good question. Yeah. I'm always a little afraid when I fly. That's what makes me so damn good. I've seen pilots who weren't afraid of anything, who would forget about checking their instruments, who flew by instinct as though they were immortal. I've pissed on the graves of those poor bastards too. The pilot who isn't a little bit afraid always screws up and when you screw up bad in a jet, you get a corporal playing taps at the expense of the government."

—LIEUTENANT COLONEL BULL MEECHAM, USMC
**in Pat Conroy's book, *The Great Santini***
**1976**

A pilot lives in a world of perfection, or not at all.

—RICHARD S. DRURY
***My Secret War***
**1979**

Between the amateur and the professional . . . there is a difference not only in degree but in kind. The skillful man is, within the function of his skill, a different psychological organization. . . . A tennis player or a watchmaker or an airplane pilot is an automatism but he is also criticism and wisdom.

—Bernard De Voto

For all professional pilots there exists a kind of guild, without charter and without by-laws. It demands no requirements for inclusion save an understanding of the wind, the compass, the rudder, and fair fellowship.

—Beryl Markham
*West with the Night*
1942

I flew in combat in Vietnam. I got shot at, I shot back, I got shot down. Compared to this flight, I felt a lot safer in combat.

—Dick Rutan
*Newsweek* magazine
*regards engine failure over the Pacific during his record round-the-world-without-refueling flight*
5 January 1987

As we went through mach one, the nose started dropping, so we just cranked that horizontal stabilizer down to keep the nose up. We got it above mach one, and once we got it above the speed of sound, then you have supersonic flow over the whole airplane, so you have no more shock waves on it that are causing buffeting. . .You really don't think about the outcome of any kind of a flight, whether it's combat, or any other kinds of flights, because you really have no control over it. . . You concentrate on what you are doing, to do the best job you can, to stay out of serious situations. And that's the way the X-1 was.

—General Charles "Chuck" Yeager
*regards the first supersonic flight, interview*
1 February, 1991

The first flight was relatively uneventful. Just one emergency, and another minor problem. A canopy-unsafe light illuminated at Mach 1.2 on the way to 1.5 at 50,000 feet, and later, during a fly-by requested by Johnson, fuel siphoning occurred. Not bad, as initial test flights go.

—ROBERT J GILLILAND
*regards the first flight of the SR-71 Blackbird*
**22 December 1964**

The hard, inescapable reality is that anyone who flies may die in an airplane.

—STEPHEN COONTS

A little mountain will kill you just as dead as a big one if you fly into it.

—STEPHEN COONTS

A pilot who says he has never been frightened in an airplane is, I'm afraid, lying.

—LOUISE THADEN

# Safety

During this period Steen and Fox were killed trying a single-engine instrument approach at Moline. Then Campbell and Leatherman hit a ridge near Elko, Nevada. In both incidents the official verdict was "Pilot error," but since their passengers, who were innocent of the controls, also failed to survive, it seemed that fate was the hunter. As it had been and would be.

—Ernest K. Gann,
***Fate is the Hunter***
**1961**

The R-101 is as safe as a house, except for the millionth chance.

—Lord Christopher Thomson
**Secretary of State for Air**
*shortly before boarding the doomed airship headed to India on its first real proving flight, The day before he had made his will*
**4 October 1930**

. . . the back motors of the ship are just holding it just enough to keep it from—it's burst into flame! It burst into flame and it's falling, it's fire, watch it, watch it, get out of the way, get out of the way, get this Charley, get this Charley, it's fire and it's rising, it's rising terrible, oh my god what do I see? it's burning—bursting into flame, and it's falling on the mooring mast and all of the folks agree that this is terrible, this is one of the worst catastrophes in the world, ohh the flames are rising, oh, four or five hundred feet into the sky. It's a terrific crash ladies and gentlemen, the smoke and it's flames now and the frame is crashing to the ground, not quite to the mooring mast, all the humanity, and all the passengers.

Screaming around me, I'm so—I can't even talk, the people, it's not fair, it's—it's—oh! I can't talk, ladies and gentleman, honest, it's a flaming mass of smoking wreckage, and everybody can hardly breathe. . . I'm concentrating. Lady, I'm sorry, honestly, I can hardly breathe, I'm going to step inside where I cannot see it. Charley that's terrible. I, I can't. . . listen folks I'm going to have to stop for a minute, just because I've lost my voice, this is the worst thing I've ever witnessed.

—HERB MORRISON,
**reporting for WLS radio**
*regards the end of LZ-129 the Hindenburg*
**6 May 1937**
*After Morrison recovered from the initial shock of the tragedy, he went on to calmly describe what he had witnessed. Listeners in Chicago and across the country didn't hear Morrison's coverage of the disaster until the next day because his report wasn't broadcast live from Lakehurst. He and engineer Charlie Nehlson had been experimenting with field recordings on huge acetate discs. They realized the gravity of their recordings as they found themselves being followed by German SS officers. After hiding out for a few hours, the two managed to make a clean getaway and get back across the country to WLS. The chilling account aired the next day on the station and was the first recorded radio news report to be broadcast nationally by NBC*

Trouble in the air is very rare. It is hitting the ground that causes it.

—AMELIA EARHART
*20 Hrs 40 Mins*
**1928**

Remember one thing, the Pk (Probability of kill) of the ground is always 100%.

—ORIGIN UNKNOWN

*but attributed to the instructors at the Naval Strike and Air Warfare Center (NSAWC) in Fallon, Nevada*

No nation can advance unless the old ideals of exploration and adventure are lived. There must be lives lost in flying, as in every other step of progress, and as many lives have been lost in the past, but there is no need to run foolish risks. The search for adventure need not entail foolhardiness. Fear is a tonic and danger should be something of a stimulant.

—LADY SOPHIE MARY HEATH

When you have two engines, you have two engines that can fall to bits. When you have four, you have four that can fall to bits. The less engines you have, the safer you are.

—FRANK FICKEISEN

**Chief Engineer for Boeing**

*replying to a complaint made by the American Airlines Allied Pilots' Association about the dangers of flying two-engine airplanes across the Pacific*

Straying off course is not recognized as a capital crime by civilized nations.

—U.S. AMBASSADOR TO THE U.N. JEANE KIRKPATRICK

*in reference to the Soviet destruction of Korean Airways Flight 007*

Tawakalt ala Allah. (I rely on God.)

—GAMIL EL-BATOUTI

**EgyptAir Flight 990 co-pilot**

**source NTSB**

*he repeated this phrase nine times while shutting off the Boeing 767's engines and pushing the jet into a fatal dive*

**31 October 1999**

What's happening, Gamil? What's happening? What is this? What is this? Did you shut the engines? Pull. Pull with me. Pull with me. Pull with me.

—CAPTAIN AHMED MAHMOUD EL HABASHY
EgyptAir flight 990
source NTSB, 13:50 EST 31
*last words on returning to the cockpit*
October 1999

There is nothing on the cockpit voice recorder or the flight data recorder to indicate that Flight 990 was intentionally crashed into the ocean.

—SHAKER KELADA, EGYPTAIR'S VICE PRESIDENT FOR SAFETY
reported in *Al-Ahram Weekly*, Cairo
17 August 2000

It's better to miss the lead story at 6 . . . than to become the lead story at 11.

—BRUCE ERION
President of the National Broadcast Pilots Assn.
1999

ATTENTION! Aircraft Designers, Operatiors, Airmen, Managers. Anxiety never disappears in a human being in an airplane—it merely remains dormant when there is no cause to arouse it. Our challenge is to keep it forever dormant.

—HAROLD HARRIS
Vice President, Pan American World Airways
circa 1950

All of the people involved in the program, to my knowledge, felt "Challenger" was quite ready to go and I made the decision, along with the recommendation of the team supporting me, that we launched.

—JESSE W. MOORE
NASA associate administrator for space flight
reported in the *New York Times*
29 January 1986

All of a sudden, space isn't friendly. All of a sudden, it's a place where people can die. . . . Many more people are going to die. But we can't explore space if the requirement is that there be no casualties; we can't do anything if the requirement is that there be no casualties.

—Isaac Asimov
on CBS television show *48 Hours*
*regards the Challenger investigation*
21 April 1988

Only realistic flight schedules should be proposed, schedules that have a reasonable chance of being met. If in this way the government would not support them, then so be it. NASA owes it to the citizens from whom it asks support to be frank, honest, and informative. . . . For a successful technology, reality must take precedence over public relations, for nature cannot be fooled.

—Richard P. Feynman
"personal observations" in an appendix to the official Challenger accident report

It is clear to us all that a tyre burst alone should never cause a loss of a public-transport aircraft.

—Sir Malcolm Field
Head of Britain's Civil Aviation Authority
*regards the Concorde*
16 August 2000

# Last Words

Portland Tower, United 173, Mayday! We're . . . the engines are flaming out — we're going down. We're not going to be able to make the airport.

—First Officer Rodrick Beebe

*The DC-8 ran out of fuel in an accident that started the assertiveness training that is now part of Crew Resource Management (CRM).*

28 December 1978

American 191 under way.

—Captain Walter Lux, American Airlines

*Last recorded words*

25 May 1979

American 191, do you want to come back? If so, what runway do you want?

—ORD tower controller

*After seeing the entire left engine and pylon of the DC-10 come off at rotation, the crippled plane crashed 30 seconds later*

25 May 1979

Reverser's deployed!

—First Officer Josef Thurner, Air Lauda 004

*Last recorded words during "impossible" in-flight deployment of the B-767's thrust reverser*

26 May 1991

Why is it turning . . . Yes it is.

—El'dar Kudrinsky, the 15-year-old son of the captain of Aeroflot 593

*He had earlier asked, "May I turn this [the control wheel] a bit," and had disconnected part of the autopilot, an action unnoticed by the adult flight crew. Last recorded words*

22 March 1994

Critter five-ninety-two, we need the, uh, closest airport available.

—First Officer Richard Hazen, ValuJet 592

*Last recorded words before crashing into the Everglades due to in-flight fire*

11 May 1996

Too late. No time, no.

—CAPTAIN CHRISTIAN MARTY, AIR FRANCE 4590 CONCORDE

*Last recorded words. ATC had just warned, "Concorde zero ... 4590, You have flames. You have flames behind you."*

**25 July 2000**

I see water and buildings . . . Oh my God! Oh my God.

—MADELINE AMY SWEENEY, AMERICAN AIRLINES FLIGHT ATTENDANT

*End of her phone call to supervisor Michael Woodward describing the hijacking of AA Flight 11. She provided many important details before the plane was crashed into the World Trade Center*

**11 September 2001**

They're coming.

—VOICE IN ARABIC

*Reportedly toward the to end of the United Flight 93 cockpit voice recorder (CVR)*

**11 September 2001**

Hang onto it. Hang onto it.

—— CAPTAIN EDWARD STATES

*Last words on the CVR, American Airlines Flight 587, from New York's JFK to the Dominican Republic. 09:15 Eastern*

**12 November 2001**

# Index

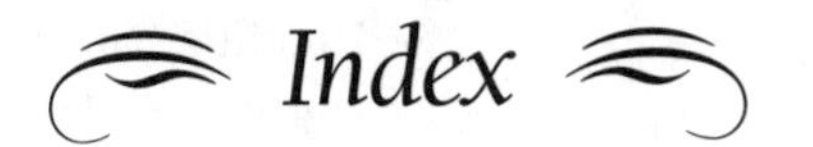

A-10 Pilot, 67
Acton, Loren, 123, 124
Aesop, 157
*Airplane!*, 85
*Airplane II, The Sequel*, 85
Albertazzie, Ralph, 107
Aleksandrov, Aleksandr, 21
Allen, Joseph, 124
Alter, Jonathan, 124
Armstrong, Neil, 19, 128
Ashby, Jeffrey S., 52
Asimov, Isaac, 43, 121, 171
Atta, Mohamed, 70
Audubon, John James, 143
Auriol, Jacqueline, 13
*Aviation Week & Space Technology*, 42

Bach, Richard, 15, 26, 59, 114, 157
Bache, Sarah, 143
Bader, Douglas, 159
Bair, Michael B., 102
Baker, David L., 157
Baker, Robert A., 135
Baker, Russell, 121
Ball, Albert, 62
Barnes, Pancho, 16
Barry, Dave, 86
Bats, Willie, 157
Batten, Jean, 14
Baturin, Yuri, 127
Baty, Peggy, 50
Bauer, Harry, 19
Bayly, Thomas Haynes, 104
Beach, Berton, 112
Beamer, Todd, 71
Bean, Alan, 21
Beaton, Cecil, 9, 111
Beaty, David, 9
Beebe, Roderick, 175
Beeson, Duane W., 55
Bell, Alexander Graham, 156
Bennett, Layton A., 156
Bertorelli, Paul, 103
Bethune, Gordon, 75, 79
Blake, William, 143
Boeing Magazine, 106
Boeing, William, 101
Boelcke, Oswald, 62
Bonaparte, Napoleon, 30
Bondar, Roberta, 124
Bracken, Earl, 60
Bradbury, Ray, 127
Brahe, Tycho, 117
Brande, Dorothea, 106
Breslau, Karen, 88
British Admiralty, 34
Brown, Arthur Whitten, 105
Bruno, Harry, 38
Buffett, Warren, 75
Burnett, Thomas E. Jr., 70
Burroughs, John, 127
Burroughs, William, 143
Burton, Richard, 22
Bush, George, 43, 80
Bush, George W., 71

Campbell, Naomi, 79
Campbell, Pete, 159
Candy, John, 111
Caras, Roger, 144
Carlin, George, 88
Carpenter, Scott, 119
Carson, Johnny, 110
Carter, Jimmy, 133

Carty, Donald, 77, 78
Cayley, George, 30
Cernan, Eugene, 128, 135
Chaffee, Roger B, 126
Charles, Jacques Alexandre Cesare, 11
Chop, Albert M., 135
Christlein, Daniel, 129
Churchill, Winston, 67, 69, 100
Clarke, Arthur C., 42, 119, 136
Cleveland, F. A., 43
Clinton, Bill, 136
Clostermann, Pierre, 13
Coates, Florence Earle, 108
Cochran, Jackie, 18, 48
Cockerell, Christopher, 100
Coffee, Jerry, 19
Coleman, Bessie, 47
Collins, Eileen, 51, 52
Collins, Michael, 6, 126
Compton, William, 77
Conrad, Pete, 119
Conroy, Pat, 161
Coonts, Stephen, 16, 160, 163
Cooper, L. Gordon "Gordo", 16, 23, 118, 134, 135
Cousins, Norman, 126
Covey, Richard, 122
Crandall, Robert, 78
Cranston, Alan, 76
Crawford, Robert, 112
Creelman, James, 107
Crippen, Bob, 124
Crocco, G. A., 38
Cromwell, Robert, 69
Cunningham, Randy "Duke", 83
Curtiss, Glenn H, 100

da Vinci, Leonardo, 7
Davies, Rick, 161
de Houthulst, Willy Omer Francois Jean Coppens, 63
De Voto, Bernard, 162
Depew, Dick, 84
Deroche, Elise, 47
Dickinson, Emily, 144
Dietrich, Marlene, 135
Doolittle, James H., 68, 106, 160
Dornberger, Dr. Walter Robert, 20
Dos Passos, John, 53
Dowding, Hugh, 134
Drury, Richard S., 161
Dryden, Hugh L., 117
du Bois, William Pene, 149
Dudgeon, A.G., 17
Dyer, Geoff, 24

Earhart, Amelia, 3, 10, 18, 47, 49, 155, 168
Eckener, Dr Hugo, 37
Eckland, K. O., 26
Edison, Thomas, 33
Edwards, Gareth, 78
Eisenhower, Dwight D., 70
El-Batouti, Gamil, 169
Epictetus, 155
Erion, Bruce, 170

Faull, Walter R. "Dick", 159
Feldman, Arlene, 49
Ferris, Richard, 76
Feynman, Richard, 133, 135, 171
Fickeisen, Frank, 169
Field, Malcolm, 171
Firth, Ned, 60
Fonck, René Paul, 63
Ford, Henry, 38
Forsyth, Frederick, 10
Fort, Cornelia, 48
Franklin, Benjamin, 143
Frontinus, 29
Fulmer, Doug, 128

Gagarin, Yuri A., 118
Galland, Adolf, 15, 68, 69
Gann, Ernest K., 18, 85, 103, 104, 167
Gates, Bill, 110
Gentile, Don S., 13
Gibson, Robert, 4
Gilliland, Robert J, 163
Gilson, Case, 133

Glenn, John, 119
Glick, Jeremy, 71
Goddard, Esther, 122
Goddard, Robert, 37
Godfrey, John T., 56
Goering, Hermann, 38, 67
Goldwater, Barry M., 24
Gorbachev, Mikhail, 136
Gould, Bruce, 24
Gower, Pauline, 160
Gray, Thomas, 30
Griffith, Del, 111

Habashy, Ahmed Mahmoud El, 170
Hall, Donald, 101
Hamill, Ed, 161
Handey, Jack, 89
Harding, Mike, 109
Harris, Anson, 80
Harris, Arthur "Bomber", 68
Harris, Harold, 170
Hartmann, Erich "Bubi", 56
Hawking, Stephen, 11
Hazen, Richard, 175
Healey, Denis, 76
Heath, Sophie Mary, 104, 169
Heinemann, Ed, 101
Heller, Joseph, 58
Hemingway, Ernest, 26, 139
Henault, Ray, 61
Herman, Richard Jr., 156
Hilton, W. F., 40
Hoffer, Eric, 120
Hoffman, Jeffrey, 122
Holland, K. G., 83
House, Dan, 83
Howard, Bevo, 157
Hubbard, Earl, 126
Hughes, Howard, 109
Hughes, Sam, 36
Hugo, Victor, 17, 32

Irwin, James B., 9

Jackman, W. J., 35
Jackson, Rodney H., 155
Jeppesen, Elrey B., 103
jetBlue airlines spokesman, 80
Johnson, Amy, 25, 105, 160
Johnson, James W., 60
Johnson, Lyndon Baines, 41, 43
Johnson, Samuel, 149
Jones, Brian, 15, 107
Jones, Harold Spencer, 40
Jones, Ira "Taffy," 57, 63
Jong, Erica, 78
Joubert, Joseph, 120

Kaempffert, Waldemar, 20
Kahn, Alfred, 77
Kai-Shek, Chiang, 70
Kaye, Jennifer, 51
Keenleyside, Hugh L., 111
Kelada, Shaker, 170
Kelleher, Herb, 79
Kelvin, Lord, 33
Kendall, Joe, 77
Keneally, Thomas, 110
Kennedy, John F., 41
Kepler, Johannes, 29
Kirkpatrick, Jeane, 169
Kissinger, Henry, 87
Knauth, Percy, 8
Knievel, Evel, 156
Koestler, Arthur, 120
Kong, T. J. "King", 83, 108
Krizek, Tom, 110
Kudrinsky, El'dar, 175
Kurtz, Walter E., 60

LaBrode, Richard, 44
Lace, Lloyd, 84
Lana de Terzi, Francesco de, 30
Lancaster, Ray, 89
Land, Robert, 79
Landis, James M., 76
Langewiesche, William, 6, 158
Langewiesche, Wolfgang, 6
Lanier, Sidney, 7
Larson, Gary, 89

Lasser, David, 37
Law, Ruth, 47
Le Conte, Joseph, 33
*Le Figaro,* 76, 151
Lee, Lincoln, 23
Leonov, Aleksei, 123
Letterman, David, 111
Lewis, Cecil, 4, 22
Lilienthal, Otto, 101
*L'Illustration,* 63
Lindbergh, Anne Morrow, 4, 9, 10, 11
Lindbergh, Charles A., 4, 11, 20, 21, 104, 112, 145, 156
Lister, Dave, 88
Littlefield, R. M., 57
Livingston, Robert, 85, 89, 159
Livni, Gidi, 55
Loening, Grover, 39
Lovell, James, 59, 133
Lowenstein, Peter, 111
Lucas, George, 117
Lux, Walter, 175
Lyall, Gavin, 6

MacDonald, George, 13, 156
MacLeish, Archibald, 9
Magellan, Ferdinand, 22
Mannin, Ethel, 69
Manyak, Louie, 84
Markgraf, Dick, 85
Markham, Beryl, 12, 158, 162
Marty, Christian, 176
May, Richard H., 55
May, Wilfred Reid "Wop," 63
McAuliffe, Christa, 51, 126
McCudden, James, 58
McCulley, Mike, 123
Meecham, Bull, 161
Melville, Herman, 24
Merbold, Ulf, 125, 126
Meyer, John C., 56
Michel, Carl, 75
Milken, Michael, 77
Miller, Arthur, 87
Miller, Austin "Dusty", 105
Miller, Joaquin, 58
Milton, John, 117
Mitchell, Edgar, 123, 125, 134
Mitchell, William "Billy", 59
Moffet, W. A., 69
Monroney, A.S. "Mike", 102
Moore, Jesse W., 170
Moore, John, 87
Morgan, Len, 103
Morris, Jeff "The Cat", 110
Morrison, Herb, 168
Mosley, Stanley, 34
Moulton, Dr. F. R., 37
Moulton, Forest Ray, 37
Murchie, Guy, Jr., 144
Murdock, Buck, 89
Musabayev, Talgat, 128

Nabokov, Vladimir, 120
Nance, John J., 9
Neeleman, David G., 75
Nesbitt, Stephen A, 122
Newcomb, Simon, 33
*New York Times,* The, 36, 125
Nishizawa, Yuji, 70
Nixon, Richard M., 114

O'Brian, William, 55
Ohnishi, Vice Admiral, 109
Olds, Robin, 61
Olson, Barbara, 70
Orlebar, Christopher, 75
*Outing,* 34, 100

Parfit, Michael, 25
Pelli, Cesar, 3
Petzinger, Thomas, Jr., 79
Picard, Jean-Luc, 85
Picasso, Pablo, 120
Piccard, Bertrand, 15, 151
*Popular Science,* 33
Potter, Dennis, 80
Potts, Michael, 101
Pound, Miles, 60
Purcell, Charles W., 139

Putnam, David Endicott, 63
Pyle, Ernie, 89

Quayle, Dan, 110
Quimby, Harriet, 47, 49, 50

Raleigh, Walter Alexander, 17
Rall, Guenther, 56
Reagan, Ronald, 24, 122, 127
Renoir, Pierre Auguste, 100
Rickenbacker, Edward "Eddie", 36, 57, 58, 135
Ridenour, Louis N., 39
Ridley, Jack, 106
Riedel, Walther, 134
Riesenberg, Felix, 155
Rolls Royce training manual, 111
Roosevelt, Eleanor, 48
Roosevelt, Theodore, 59
Rumsfeld, Donald H, 71
Russell, Thomas, 34
Rutan, Dick, 107, 162
Rutherford, Ernest, 67

Safire, William, 114
Sagan, Carl, 42, 44, 129
Saint-ExupÈry, Antoine de, 10, 12, 23, 24, 25
Samson, Charles Rumney, 16, 158
Sandburg, Carl, 106
Santos-Dumont, Alberto, 4, 149, 150
Sayer, Gerald, 38
Schweickart, Russell, 8, 125
*Scientific American,* 31
Scobee, Dick, 122
Scott, David R., 125
Scott, Robert L., Jr., 7
Sevastyanov, Vitali, 123
Shakespeare, William, 3, 144
Shatner, William, 117
Shaw, George Bernard, 13, 83
Shepard, Alan B. Jr., 118, 119
Shugrue, Martin R. Jr., 78
Sideshow Bob, 111
Sikorsky, Igor, 39, 40
Skelton, Betty, 50
Slattery, Paul, 88
Slavsky, E. P., 40
Slayton, Donald "Deke", 118
Smith, C. R., 77
Smith, Dean, 84
Smith, Edward J., 156
Snow, C. P., 121
Socrates, 3
Sondheim, Stephen, 80
Sonnenfeld, Barry, 83
Soucy, Colin, 110
States, Edward, 176
Steiner, John E. "Jack", 101
Stevens, Bob, 87
Stevenson, Robert Louis, 17, 150
Still, Susan, 51
Stinson, Katherine, 48, 100
Summerfield, Arthur E., 41
*Sun,* The, 87
Sweeney, Madeline Amy, 176

Tennekes, H., 76
Terry, Ron, 89
Thaden, Louise, 48, 49, 163
Thomson, Christopher, 167
Thoreau, Henry David, 67, 109, 144
Thorp, Rex, 158
Thurner, Josef, 175
Tiburzi, Bonnie, 50
*Times* of London, The, 34
Tito, Dennis, 127
*Top Gun,* 108
Townsend, Peter, 57
Tracy, Spencer, 100
Treacy, Michael, 84
Trenchard, Hugh M., 67
Truman, Harry S., 134
Tseu, P'ao-Pou, 3
Tsiokovsky, Konstantin E, 32
Twain, Mark, 3
Twining, Nathan F., 67

Udet, Ernst, 61
Udvar-Hazy, Steven, 105

van Buren, Martin, 30
van der Riet Wooley, Richard, 38, 40
van Gogh, Vincent, 20
Vedrines, Jules, 35
Verne, Jules, 31
Vidal, Gore, 43
von Braun, Wernher, 40, 42, 117, 129
von Karman, Theodore, 101
von Richthofer, Manfred, 62
Vonnegut, Kurt, Jr., 120

Wachhorst, Wyn, 26
Walker, Charles, 123
Wallace, Lane, 3, 16
Walpole, Horace, 149
Wang, Taylor, 126
Weiner, Earl, 159
Welles, Orson, 84
Wells, H. G., 35
Welsh, Edward O., 41
Westbrook, Robert B. "Westy", 57
Westcott, E. N., 61
White, Frank, 21
White, T. H., 25
Wikert, Alinda, 50
Wilcox, Robin, 139
Wilde, Oscar, 17
Wilkins, John, 29
Williams, Donald, 122
Wilson, Gill Robb, 7, 114
Wilson, Ned, 76
Windstaff, W.W., 159
Wise, Leon M., 84
Wittenborn, John, 88
Wood, Kingsley, 68
Wordsworth, William, 150
Wren, Christopher, 29
Wright, Orville, 15, 34, 99, 101
Wright, Wilbur, 4, 34, 155, 157

Yeager, Charles "Chuck", 102, 160, 162

## About the Author

**Dave English** is an airline captain with an American air carrier. A resident of Milwaukee, Dave is also a contributing editor to *Airways* magazine and has published articles in *Air Line Pilot*, *Airliners*, *Aviation Consumer*, *Flight Training*, *IFR*, and *Professional Pilot* magazines. He is the author of McGraw-Hill's acclaimed best-seller *Slipping the Surly Bonds: Great Quotations on Flight*.